Patrick Moore's Practical Astronomy Series

D1174718

DATE DUE

For other titles published in the series, go to
www.springer.com
click on the series discipline
click on the heading "Series"
click on the name of the series

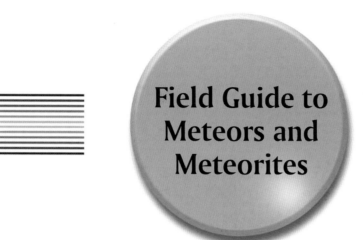

Field Guide to Meteors and Meteorites

O. Richard Norton
Lawrence A. Chitwood

 Springer

ISBN 978-1-84800-156-5   e-ISBN 978-1-84800-157-2
DOI: 10.1007/978-1-84800-157-2

British Library Cataloguing in Publication Data
A catalogue record for this book is available from the British Library

Library of Congress Control Number: 2008921357

Patrick Moore's Practical Astronomy Series ISSM: 1617-7185

Cover illustration: Photo of Hammada al Hamra 335 found in Libya in 2004. Courtesy of Dr. Svend Buhl, www.meteorite-recon.com.

Printed on acid-free paper

9 8 7

Springer Science + Business Media
springer.com

To Lawrence A. Chitwood,
co-author, colleague and good friend,
who died suddenly shortly after this book was completed -
His passion for minerals in meteorites, expertise in the use of the petrographic microscope,
and remarkable ability to make this complex science understandable
contributed greatly to the production of this book.
He will be sorely missed.

- O. Richard Norton

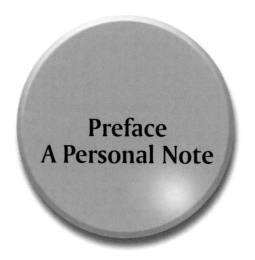

Preface
A Personal Note

When I was a 13-year-old kid nearly six decades ago, I discovered the heavens. That discovery changed everything. Finding my way around the night time sky was the first challenge. As I learned the celestial stick figures of Greek and Roman mythology placed there as monuments to ancient gods, they became stationary signposts that allowed me to wander from figure to figure, retracing my steps from celestial pole to celestial equator. I soon learned that hidden within these constellations were objects of great beauty and wonder—star clusters, nebulae, double stars, and galaxies. It wasn't long before I laid aside the simple star charts that I had carefully cut out from the Long Beach, California *Press-Telegram*. In their stead, I purchased a book by my namesake, *Norton's Star Atlas*. This marvelous book had sky charts showing many more stars than I had learned about from the newspaper maps, stars so faint that I could barely see them with the unaided eye. There were exceptionally bright stars also. Some of them moved among the starry background, usually moving east but occasionally changing direction and moving west for several weeks before stopping and resuming their easterly courses. I learned that these were the planets of the Solar System and the motions they displayed against the stars had perplexed ancient Greek astronomers for centuries.

It wasn't long before I recognized my physical limitations. If I wanted to continue this adventure, I would need a telescope. Back in those days most amateur astronomers made their own telescopes from mirror to mount. (Nowadays, if you have the dollars you can purchase telescopes that match some of the best found in professional observatories.) I made a 4-in. reflecting telescope out of plate glass and pipe fittings. This modest telescope would be my constant companion until I graduated from high school. With it, I could explore the mountains and craters on the Moon, the rings of Saturn, Jupiter and its Moons, the polar caps of Mars, and on and on. No longer this naked eye astronomy. With my telescope, I had reached out to bring the celestial sphere with all of its wonders a bit closer to me. Besides the obvious Solar System objects, there were more subtle things that were a challenge to find and observe. Asteroids! Through my telescope, I could only see the largest and brightest as a yellowish point of light moving among the stars between Mars and Jupiter. And then there was the occasional comet that seemed to defy all the rules, crossing the orbits of the inner planets with total disregard for their stately motions. Comets were fascinating because, unlike the constellations, they could be seen to change from night to night or even hour to hour, leaving behind a diaphanous tail of gaseous matter and dust. There were some things in the sky that did move quickly—"falling stars." I remember lying on my back when I was 9 years old watching the Draconid meteor shower of October 9, 1946, and waiting for the stars to fall. I recall staring at a particularly bright star (probably Vega) and anticipating its final fall to Earth. Needless to say, it's still there among the summer stars. At that early age, I wasn't aware that falling stars (or shooting stars) were not stars at all, but dust specks from comets descending to Earth. Once I had learned this association several years later, I was primed to ask more pertinent questions. Could any of this comet dust survive passage through Earth's atmosphere? Might there be comet dust in my mother's vacuum cleaner? Comet dust particles, now more accurately called interplanetary dust particles (IDPs), have been collected in the stratosphere by sticky-winged high flying aircraft as we will see. Most of the dust particles are too small to produce visible meteors but annual meteor showers show us that larger particles, a millimeter or greater, do exist and do produce meteors, some very bright ones.

Can we go further? Is there a connection between IDPs and other objects of subplanetary mass, namely, the asteroids? Occasionally, brilliant fireballs are observed passing through Earth's atmosphere leaving a trail of gas and dust behind. These have been photographed in flight, and their orbital paths have been extrapolated backward terminating in the asteroid belt. This implies that there are still large chunks of rocky material out there that have left the asteroid belt and are on their way to Earth. Today our museums house thousands of rocky samples of asteroids that have come to us from space—free of charge. They are relics of the early Solar System. So we have moved from the fixed stars to celestial objects we can actually touch. Even rock samples collected on the Moon during the Apollo missions have their counterparts on Earth. About 50 stones of lunar origin have been found on Earth in the past decade. More discoveries will follow as meteorite hunters (scientists and amateurs) continue to comb the hot and cold deserts of the world to touch what was once beyond their reach.

It wasn't long before I began to realize that the study of the Solar System requires more than simply acquiring beautiful optical images of distant Solar System objects, as wonderful as they are. Over the years, I recognized the need to expand my knowledge of meteorites to include such diverse fields as petrology, optical mineralogy, petrography, and the study of minerals as they pertain to meteoritics, the science of meteorites. I invited Lawrence A. Chitwood, volcanologist and geologist for the Deschutes National Forest in central Oregon and an expert in igneous petrology and mineralogy, to write this book with me for he has much of the knowledge that is necessary to study meteoritics. (We met not as scientists, but as musicians. We both play the piano and have performed many classical duets together. Curiously, I learned that he also made a telescope when he was a kid and he too spent hours under the sky memorizing the constellations.)

The connection has now been made between those tiny specks of dust that flash across the night sky, their asteroidal precursors and meteorites. No longer must amateur astronomers be content with passively observing the heavens. Not only are meteorites fragments of asteroids but some also contain interstellar grains—and we can actually hold them in our hands. These space rocks have made it to Earth by the thousands, even though it took us centuries to come to terms with them, to realize their true nature. They are scattered all over Earth from pole to pole waiting to be discovered. A new world of exploration awaits the backyard scientist. Amateur astronomers, in particular, have surveyed the Solar System with their telescopes but their surveys remain incomplete. Now it is time to look down as well as up; to set aside your telescopes, and arm yourselves with metal detectors, magnets, magnifiers, and microscopes and prepare to explore the wonders of the Solar System locked inside these rocks from space for the last 4.56 billion years. Let this book be your guide to the smaller bodies of the Solar System.

O. Richard Norton

Acknowledgements

This book could not have been written without the assistance of many people. Scientists, collectors, meteorite dealers and hunters the world over responded enthusiastically to the notion of a field guide and to our requests for images of all types of meteorites. It is not possible to list all of them here, but their names appear in photo captions throughout the book and we are grateful to them all. In some cases, they supplied wonderful images of specimens we never could have acquired ourselves. In other cases, they sent specimens for us to photograph.

Special thanks must be given to John Watson, Consultant Publisher in London, whose initial support of this project led to the idea of a Field Guide to Meteors and Meteorites. We must also thank our Editor Harry Blom for insisting that this book would be full color, thereby allowing us to present the many wonderful images of these elusive visitors from space. Christopher Coughlin, Associate Editor, and Joseph Quatela, Production Editor worked to ensure that the book would fulfill its stated purpose and would function as a general introduction to the subject and a field guide to specific meteorite types.

Lastly and certainly not least of all we must thank our wives. The book, as often is the case, took longer than expected to write and they were supportive from beginning to end. Karen Chitwood read and reread much of the text and made many helpful suggestions. Dorothy Sigler Norton, a scientific illustrator, contributed most of the diagrams and other illustrations, as well as online research, editing and endless correspondence.

Contents

Contents

Appendices

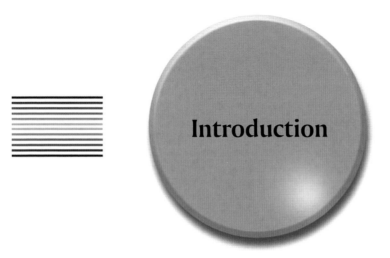

Introduction

Popular and scientific interest in meteorites is at an all-time high and growing. Weekend meteorite hunters armed with magnets and metal detectors have successfully scoured the dry deserts of the western United States. Serious collectors and dealers have added immeasurably to the number and availability of meteorites in recent years. Many new meteorites have come from the hot deserts of northern Africa, and researchers have recovered large numbers from the ice fields of Antarctica.

This book fulfills a need for a well-illustrated identification guide for amateur meteorite hunters and collectors, amateur astronomers, and anyone interested in the great space explorations and discoveries of our day.

The title of this book—*A Field Guide to Meteors and Meteorites*—suggests that one can walk around with this book, find a meteorite, and quickly identify what kind it is. After all, a field guide to birds or trees often requires only a brief comparison between illustrations in the guide and sharp-eyed observations in the field to make an identification. A field guide to rocks and minerals, on the other hand, relies heavily on close-up details and physical tests. And so does this book. After all, the natural fragments that shower our atmosphere are rocks and minerals from space. Those that survive the intense heat and forces of entry fall to Earth and disguise themselves as Earth rocks until someone using this book learns to unmask them as meteorites.

This field guide is divided into three parts: Ancient Fragments of the Solar System, The Family of Meteorites, and Collecting and Analyzing Meteorites. The three chapters of the first part describe the types, origins, and observations of these fragments. They include discussions of interplanetary dust particles, zodiacal light, meteors and meteor storms, observing and photographing meteors, the asteroid–meteorite connection, and surface features of meteorites.

The second part, the centerpiece of this book, is a lavishly illustrated guide to the identification, classification, and petrography of meteorites. It includes their external appearance, appearance in cut sections, and microscopic features. Five chapters describe and illustrate the chondrites, primitive and asteroidal achondrites, planetary and lunar achondrites, iron meteorites, and stony-iron meteorites. The beauty and variety of chondrules is highlighted in a colorful chondrule gallery. Another photographic gallery illustrates a few of the most commonly found objects: meteorwrongs.

The third part discusses collecting and analyzing meteorites. The first of two chapters answers important questions about collecting meteorites in the field: Where is the best place to search? What should a freshly fallen meteorite look like? The second chapter presents a user's guide to the petrographic microscope. This unique microscope remains vitally important to the understanding and classification of meteorites. And it brings vibrant colors to otherwise drab-looking meteorites. This chapter provides an illustrated guide to the building blocks of meteorites, illustrations of textures, and classification tables for chondrules, chondritic meteorites, petrographic type, shock grade, and weathering grade. Helpful hints are given for how to make your own simple petrographic microscope and how to take photographs through the microscope.

Now it's your turn. Whether you are hunting them in the field or studying them in your home laboratory, we invite you to begin your adventure with meteorites, these enigmatic remnants of the early Solar System. The tools are here. The rest is up to you.

Part I

Ancient Fragments of the Solar System

CHAPTER ONE

Interplanetary Dust and Meteors

Along the plane of the Milky Way Galaxy are thousands of enormous massive clouds of gas and dust. Typically 10–50 parsecs across, they are *dark molecular clouds* made of hydrogen and helium, organic molecules and dust. Cold and dark, they are so dense that light cannot easily penetrate them. We see these interstellar clouds as silhouettes against the light of background stars or stars immersed in bright diffuse hydrogen clouds. The dust particles are composed of heavy elements like iron and carbon along with simple molecules such as silicon carbide, graphite, and diamond. This dust is scattered among silicate grains forming cores around which ice deposits. Collectively referred to as *interstellar dust particles*, they were first found in primitive carbonaceous chondrite meteorites. These heavier components are contaminants, made not in the cloud itself but manufactured in nearby massive stars through thermonuclear processes in their interiors and then scattered throughout the Milky Way Galaxy by strong stellar winds. The molecular clouds often fragment due to instabilities within the clouds or perhaps because of massive explosions of nearby supernovae that generate enormous shock waves capable of collapsing and fragmenting them. As the fragments collapse they form cores of much higher density, about 10^4 molecules/cm^3. They continue to collapse until a rapidly rotating protostar, or protoSun, forms surrounded by a thick disk of gas and dust along the protoSun's equator. This is the protoplanetary accretion disk. Soon after the disk forms and begins to cool, accretion of the first refractory minerals (stable at high temperatures) begins to take place. All of this occurred within a few 100,000 years of our Solar System's history. The developing protoSun and accompanying disk remained hidden within a cocoon-like core, the stellar nursery. Amateur and professional astronomers are quite familiar with such nebulae, having for years imaged them in their telescopes. A well-known region is the Scutum star cloud that passes through the constellation of Scutum in the summer Milky Way. Amateur telescopes reveal an open star cluster composed of hot type O and B stars enveloped in nebulosity, marked on star charts as Messier 16. Several years ago the central part, called the Eagle Nebula, was imaged by the Hubble Space Telescope, revealing huge columns of dark molecular clouds set against bright diffuse hydrogen regions (Figure 1.1). Interstellar dust grains scatter the blue light of the hot embedded stars. At the tips of these columns photoevaporation is taking place, revealing the position of emerging new solar systems.

Figure 1.1. A Hubble Space Telescope image of the top of a column of molecular hydrogen and dust thought to be an incubator of newly formed stars in M16, the Eagle Nebula. Each projection is slightly larger than our Solar System. Courtesy of NASA, ESA, and STScI (J. Hester and P. Scowen, Arizona State University).

Interplanetary Dust Particles (IDPs)

Today, the Solar System is still a dusty place but the dust once occupying the accretion disk surrounding the protoSun is not the same dust we find today throughout the inner Solar System. We know that interplanetary dust particles (IDPs) or *micrometeoroids* come primarily from two sources related to the Solar System: comets and asteroids. A third source of tiny particles takes astronomers well beyond the Solar System and into the interstellar environment of the Milky Way Galaxy light years beyond the Sun. These are the interstellar particles briefly mentioned earlier composed primarily of carbon-based minerals whose origins are among the Galaxy's supernovae remnants.

Lifetime of IDPs

What fate IDPs suffer in their lifetime in the Solar System depends upon their mass and size. IDPs are the smallest members of the Solar System. They range in size from about a micrometer (10^{-4} cm) to several micrometers. Particles in this size range are slowed high in Earth's atmosphere before they have had a chance to heat up to produce visible meteors. Their mass to volume ratio is small meaning they have large surface areas relative to their masses, which allows them to radiate away the heat of atmospheric entry rapidly enough to remain cool and relatively pristine. Given an entry velocity of Earth's escape velocity of 7.7 mi/s (11.7 km/s), for entry without ablation the critical size limit for such particles is about 52 μm in diameter. These dust specks eventually settle to the ground as micrometeorites. Particles less than one micrometer in size are subject to the Sun's radiation pressure and the pressure of the solar wind. The two working together produce a continuous radial force that eventually drives the particles out of the Solar System. Early spacecraft such as Pioneer 10 and 11 venturing outward away from the Sun encountered these small particles out to about 18 AU (Earth's mean distance from the Sun).

Particles from a few micrometers to several millimeters diameter experience a different fate. For these larger particles the Sun's gravitational force exceeds the repulsive force of the Sun's radiation pressure. The gravitational force sends them into elliptical orbits around the Sun. The direction of the force of radiation pressure is not exactly radial with the Sun but rather has a small component in a direction opposite to that of the particle's motion around the Sun. As an analogy, this displacement is experienced when driving through a rain storm. The vertical raindrops have a horizontal component of motion, that is, a velocity equal and opposite to the car's motion. For particles orbiting the Sun, photons do not appear to come from an exact radial direction away from the Sun but have a slight velocity component in a direction opposite to that of the particle's motion. Thus, radiation pressure produces a slight backward force on them. This force acts like a drag on the orbital motion of the particle, changing its orbit from an ellipse to a circle. The particle's orbit gradually diminishes in size until it ultimately spirals into the Sun. The time scale for its fall into the Sun depends on the original position of the particle (its orbital eccentricity and perihelion distance). For example, a particle one millimeter in diameter originating in the asteroid belt 2.8 AU from the Sun will spiral into the Sun in about 60 million years. However, most of the particles are found within the inner Solar System between the Sun and Mars. A 1-mm diameter particle at a distance of 1 AU and an orbital eccentricity of 0.7 (a typical comet orbital eccentricity) will fall into the Sun much sooner, within a million years. This means that the Solar System must be rapidly losing these particles. Every second about 8 metric tons fall into the Sun. This amounts to about 250 million metric tons per year.

The Zodiacal Light

Perhaps the most extraordinary manifestation of IDPs is the *zodiacal light* (Figure 1.2). This subtle glow is visible on clear moonless nights after sundown in the Spring and before sunrise in the Fall. At that time of year the ecliptic plane (plane of the Solar System) stands more nearly vertical than at any other time of the year. It can be seen as a broad cone of light centered on the ecliptic and extending from the sunset point on the horizon (or sunrise point) to near the zenith. It often appears as bright as the Milky Way with its brightest point near the sunset or sunrise point on the horizon, gradually fading away before reaching the zenith. Given the right time of year and a dark sky, the zodiacal light can account for over 40% of the night sky glow. A spectrum of the zodiacal light shows it to be sunlight reflecting off countless billions of IDPs. A much fainter glow of light requiring very clear dark skies can sometimes be observed opposite the zodiacal light in the west. This is called the *gegenschein*, German for "counterglow." This is also sunlight reflecting off IDPs but in the opposite direction. Figure 1.3 shows the geometry of both. The zodiacal light originates between the Sun and Earth with the Sun's light forward-scattered as it passes through the dust. Here, the particles are closest to the Sun and more numerous, providing an effective scattering agent. The gegenschein is opposite the Sun with Earth in between. In this position the particle density is much less and the light must travel a greater distance to Earth. Under the best of atmospheric conditions the zodiacal light and gegenschein have been seen to connect with each other near the zenith.

So the zodiacal light is actually an enormous dust cloud that fills the inner Solar System between the terrestrial planets from inside Mercury's orbit out to Jupiter. It is not without structure, however. The Infrared Astronomical Satellite (IRAS) launched in late January, 1983, carried infrared detectors that could sense heat radiating off dust particles positioned along the ecliptic plane. This was the first in-space detection of the dust responsible for the zodiacal light. Further, IRAS detected a broad zone of dust on either side of the ecliptic plane and encircling the asteroid belt between 200 and 300 million miles from the Sun. This was the first indication that at least some

Figure 1.2. Time exposure of the zodiacal light seen about 2 h after sundown in the Spring. It appears as a faint broad elliptical cone of light extending from the sunset point on the horizon to nearly 60 degrees above the horizon.

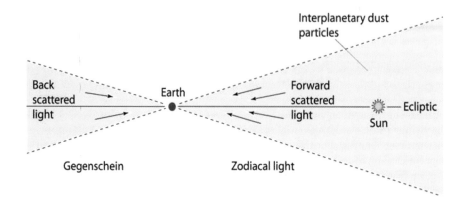

Figure 1.3. Geometry of the zodiacal light. Zodiacal light is both forward scattered and back scattered off interplanetary dust particles. Forward scattering is the brighter of the two and is seen between Earth and the Sun. Back scattered light (gegenschein) occurs beyond Earth and opposite the Sun and is much fainter.

of the zodiacal dust, perhaps as much as 40–50%, was created by dust released from colliding asteroids, leading one astronomer to exclaim that "the asteroid belt is slowly grinding itself down." In addition to this broad dust belt, IRAS detected bands of dust that corresponded with the paths of short period comets. These dust bands, referred to as *meteor streams*, are simply denser strands of cometary material left behind by comets during their perihelion passages. This material gradually spreads apart eventually becoming part of the broadly diffuse band of the zodiacal light.

Photographing the Zodiacal Light

Photographing the zodiacal light requires only a standard 35-mm film camera equipped with a fast lens (F/2.8 or faster) and a focal length short enough to image the entire light shaft which measures 15° near the horizon on either side of the ecliptic and 60° from the horizon to the vertex of the light cone. A 28-mm focal length lens can cover an angular diameter of 76° across the diagonal on a 35-mm frame. You can expose for only about 25 s at that focal length before the stars begin to show diurnal motion. Longer exposures mean that you will need to provide a clock driven telescope with the 35-mm camera mounted on the telescope tube. Hand guiding the telescope should not be necessary if you have accurately aligned the telescope onto the north celestial pole. Using 400 ISO slide or print film and exposures between 2 and 8 min should produce a good picture. Photographing the gegenschein is a much greater challenge requiring guided exposures of as long as a half hour.

Collecting IDPs—No Job for an Amateur

Comets with their dusty tails are obvious sources of IDPs. These particles are among the smallest members of the Solar System and are thought to be responsible for over 50% of the dust producing the zodiacal light. Infrared satellites have detected bands of cometary dust crossing the inner Solar System and intermingling with the dust of the zodiacal light. We saw that these dust bands were first detected by IRAS in 1983 and then later verified by the COBE (Cosmic Background Explorer Satellite) satellite. Beginning in the late 1950s, attempts were made by scientists at the Smithsonian Astrophysical Observatory to collect extraterrestrial particles. It was correctly assumed that these particles remained suspended in Earth's stratosphere for several weeks, slowly settling more or less uniformly over Earth's surface. Therefore, the first collection attempts were logically made from the surface, the ultimate resting place for the particles. As it turned out, Earth's surface was absolutely the worse place to collect interplanetary dust. No one knew what to look for in the maze of natural and man-made terrestrial contaminant. Most of the particles appeared to be of terrestrial origin. It soon became evident that collection needed to be made where terrestrial contaminants were minimal—in the stratosphere. In the late 1960s balloons were used to carry dust collectors to an altitude of as high as 23 mi (37 km) within the upper stratosphere. Even at this high altitude terrestrial contaminants were encountered and interplanetary dust particles turned out to be far fewer in number than anticipated. Moreover, balloon-carrying dust collectors were not able to sample a large enough volume of space to hope to collect adequate numbers of particles. Even when the collection was successful, in the mid 1960s there was no way to manipulate and chemically analyze the minuscule dust specks. The particles averaged about 10^{-3}-cm diameter and many were composed of subparticles between 5 and 150 nm. Today's electron and ion microprobes were still under development when these early investigations were being conducted.

By the early 1970s the situation began to change. Donald Brownlee and Paul Hodge of the University of Washington's Interplanetary Dust Laboratory began to reconsider the possibilities of dust collection, this time using high flying aircraft. Beginning with the U2 spy plane equipped with efficient dust collectors under the wings, these planes (Figure 1.4) have enormous wing spans that can support the slow flying aircraft at a 12 mi (20 km) altitude. In 1973, Brownlee and Hodge made the first successful series of test flights at 22 mi (35 km) altitude. In ten flights they successfully bagged some 300 IDPs.

In the atmosphere, taken as a whole, the lower stratosphere turned out to be the best place to collect IDPs. Many of the particles are slowed down by collisions with atmospheric molecules at altitudes between 60 and 80 mi (100–130 km). This deceleration causes a 10^6-fold increase in the flux of particles into the lower stratosphere compared to the upper, which in turn increases the collection efficiency. This is the altitude in which atmospheric heating melts many incoming micrometeoroids, producing meteors. Each particle experiences a 5–15 s flash-heating event as a function of its size, mass, entry velocity and entry angle. Asteroid dust experiences less heating than cometary dust of the same size and density. A knowledge of the particle's maximum flash-heating temperature allows the particle's entry velocity (entry into Earth's atmosphere) to

Figure 1.4. The first plane to collect IDPs from the stratosphere was the now famous U2 spy plane. Later, NASA equipped the experimental aircraft seen here, the ER-2, with dust collectors along the wings that succeeded in collecting hundreds of particles. Courtesy of NASA Dryden Flight Research Center.

be estimated. Asteroid dust enters the atmosphere at about 7 mi/s (12 km/s) while comet dust has entry velocities of about 12 mi/s (20 km/s) or even higher. Since cometary IDPs enter the atmosphere at a higher entry velocity than asteroidal IDPs, a distinction could be made between cometary and asteroidal IDPs.

Physical Properties of IDPs

While Brownlee and associates were seeing their first successes collecting IDPs in the lower stratosphere other scientists were perfecting electron and ion microprobes, instruments capable of analyzing very small particles. The electron microprobe was developed by Raymond Castaing in France in 1951. This analytical instrument gained great precision in the 1960s. One of the first highly accurate microanalyses of meteorites was made by meteoriticists Klaus Keil and Kurt Fredriksson in 1964. They considered the electron microprobe to be as important to the geosciences in the 20th century as the introduction of the polarizing petrographic microscope was to mid-nineteenth century mineralogists. The electron microprobe employs a finely focused beam of electrons that is directed onto a highly polished sample. The electron beam produces X-rays in the sample which are then separated by an X-ray spectrometer into wavelengths characteristic of the elements present. Since X-ray intensities are related to the amount of the element present in the mineral, it becomes possible to determine the elemental composition of the mineral in question. The ion microprobe operates in a similar fashion. Finely focused beams of ions are directed at the highly polished surface of a rock specimen. The ion beam converts a tiny amount of the target rock into a vaporous cloud by a process called *sputtering* which is then injected into a mass spectrometer where the elemental composition of rocks are determined.

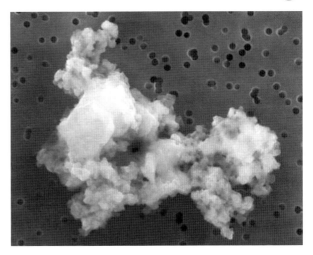

Figure 1.5. This 10-μm diameter IDP was collected by a NASA aircraft at an altitude of 20 km. It is an anhydrous porous "fluffy" IDP containing minerals similar to those found in asteroid dust and primitive meteorites but with higher carbon and volatile element abundances. Courtesy of NASA.

IDPs are among the smallest successfully collected bodies in the Solar System. Typical IDPs have "rough" diameters from 1 to 50 μm, thus the need for an electron microscope to study them. Textures range from smooth to "fluffy" in appearance. Most IDPs consist of aggregates of smaller mineral grains typically ranging in size from <0.1 to 3 μm. IDP densities vary from 0.7 to 2.2 g/cm³. The typical mass of IDPs range from 10^{-12} to 10^{-9} g. There are two basic groups of IDPs that differ in their chemistry and textures: micrometer-sized chondritic aggregates and non-chondritic particles. Chondritic IDPs have cosmic elemental abundances similar to the Sun like CI carbonaceous chondrite meteorites (see Chap. 4). Most of the chondritic IDPs are compact aggregates of closely bound minerals with little pore space. The remaining chondritic IDPs are loosely consolidated and porous (Figure 1.5). Aggregate IDPs have densities between 2 g/cm³ and 3.5 g/cm³, similar to carbonaceous meteorites. Nearly all the chondritic IDPs are dominated by either anhydrous silicates (minerals lacking water) or layer-lattice silicates (layered structure in their crystal lattices). The anhydrous silicates include single grains of iron-poor pyroxene and olivine with minor amounts of magnetite, kamacite (low-nickel iron), cohenite (iron-nickel carbide), and chromite. They are highly porous aggregates. The layer-lattice silicates include smectite (a clay mineral) as the primary mineral and serpentine. The most unusual particles called "tar balls" are composed of aggregates of tiny crystals held together, once again, by carbonaceous matter and glass.

In some respects, IDPs seem closely related to CI and CM carbonaceous chondrites. Although they contain minerals common to carbonaceous meteorites, other minerals and structures seem to be unique to IDPs alone. Most striking is that many IDPs contain 50% carbon compared to CI carbonaceous chondrites which, despite their name, contain only about 5% carbon.

Deep Sea IDPs

About two thirds of Earth's surface is covered by water so it is no surprise that the deep oceans are good catchment basins for IDPs. The search for IDPs on the ocean floor actually began over a century ago. These particles have been characterized as ablation products of iron meteorites that have melted during their passage through Earth's atmosphere. Most are spherical particles averaging about 1 mm in diameter. They are black in color and fortunately contain sufficient magnetite to allow them to be retrieved from the ocean floor with magnetic collectors. The magnetite combines with taenite (a nickel-rich iron alloy) and forms a shell of taenite around a core body of the same nickel-rich alloy.

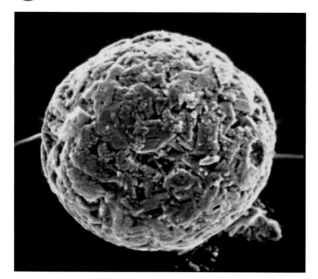

Figure 1.6. This spherical IDP was found with a magnetic collector in the middle of the Pacific Ocean at a depth of over 4000 m. The spherical shape formed when the particle melted briefly at 80-km altitude during entry, vaporizing some elements (carbon, sodium and sulfur) and leaving primarily olivine, glass and magnetite. Its elemental composition is close to that of the primitive chondrite meteorites. Courtesy of Dr. Don Brownlee, University of Washington.

More than half of the spheroids have chondritic compositions similar to type 1 carbonaceous chondrites and are thought to be ablation products of chondritic stony meteorites. The primary silicate is magnesium-rich olivine and magnetite. The olivine was melted and recrystallized during atmospheric passage and metallic iron found in chondritic meteorites was rapidly oxidized and combined with glass formed during the ablation process. The resultant material is similar to the fusion crusts that form around meteoroids during their passage through the atmosphere (Figure 1.6).

Collecting IDPs in Space—The Stardust Mission

We have seen that Earth's environment is contaminated with terrestrial dust particles from the surface through the stratosphere. To avoid terrestrial contaminant the most ideal approach is to collect them in the space environment; better yet, to collect them directly from the sources—the comets. The idea of collecting IDPs from outside the atmosphere is a concept that Donald Brownlee (University of Washington) and Peter Tsou (Jet Propulsion Laboratory) had been working on since the early 1980s. In 1994 they were awarded a NASA grant to design and build a space probe that could rendezvous with a comet, collect IDPs from the comet's dusty coma and return the trapped particles to Earth. Now, it is one level of difficulty to fly an aircraft 12 mi (19 km) into the stratosphere collecting IDPs along the way. But it is an enormous technical leap to send a spacecraft to a comet. The project was aptly called *Project Stardust*. Certainly the most technically difficult aspect of the mission was to develop the means by which cometary particles could be collected without destroying them in the process. As we saw earlier, cometary particles have much higher impact velocities than, say, a main belt asteroid particle. The particles would impact the collecting surfaces causing them to instantly vaporize on contact. Materials scientists at the Jet Propulsion Laboratory were given the daunting task of developing a material of such low density that IDPs could effectively penetrate the material slowing them down along the way and ultimately trapping them. This material, called *aerogel*, has an extraordinary low density—only $0.05 \, g/cm^3$. Aerogel's remarkable insulating properties allow the substance to transfer energy from the impacting hypervelocity particles to the aerogel, resulting in a "soft-capture" of the cometary particle imbedded within the material.

On February 7, 1999 the Stardust spacecraft was launched on its epic-making journey to the periodic Comet Wild 2 (pronounced "Vilt"). A year later Stardust began collecting interstellar dust particles. These are particles from beyond the Solar System. Scientists had looked for such interstellar particles long before the Stardust mission became a reality. Decades earlier (1978) geochemists had discovered traces of the noble gas, xenon, within the Murchison CM2 carbonaceous chondrite. Five years later they announced the discovery of two isotopes of carbon that could act as carrier grains with a crystal structure that could capture and retain the xenon gas as the carbon grains were expelled into the interstellar medium. Finally, in 1987 they presented conclusive proof that the carbon carrier was in the form of tiny nanodiamonds, diamonds a thousand times smaller than an average IDP. As predicted, xenon was found locked in the crystal structure of the interstellar diamonds. The interstellar diamonds were extracted from the Murchison meteorite by dissolving away the entire meteorite and collecting the residue containing the insoluble diamond dust. Ed Anders, one member of the team of scientists who first extracted the tiny diamonds at the University of Chicago when referring to the technique they used, posed the question, "How do you find a needle in a haystack?" His answer was succinct, "You burn down the haystack!"

On January 2, 2004, the Stardust spacecraft passed through the dust-laden coma of Comet Wild 2 and successfully captured a cargo of IDPs along the way. At the same time the nucleus of Comet Wild 2 was photographed from a distance of a mere 150 km, revealing an icy, crater-marked surface as the spacecraft performed flawlessly (Figure 1.7). After the flyby, the spacecraft's final task was to release the sample return capsule with its precious cargo. On January 15, 2006, the sample return capsule reentered Earth's atmosphere, briefly flamed as a bright meteor and successfully landed on the Utah desert floor, its protective cargo of aerogel and trapped cometary particles undamaged (Figure 1.8). When the capsule was opened it revealed dozens of cometary particles. A preliminary survey showed a two micrometer particle made up of the magnesium-rich mineral, forsterite, one of the several olivines commonly found in igneous rock on Earth, in asteroids and—you probably guessed it—in chondritic stony meteorites.

Figure 1.7. The Stardust spacecraft encountered Comet Wild 2 on January 2, 2004, and successfully collected hundreds of particles. At the same time it made this picture of the icy cratered nucleus of the comet from a distance of only 150 km. Courtesy of NASA.

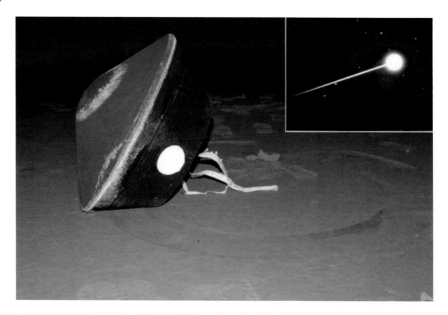

Figure 1.8. After a 7-year-long journey, the Stardust capsule reentered Earth's atmosphere, flaming briefly as it formed a comet-like body with a long narrow tail (inset). Its protective heat shield had done its job. The cone-shaped capsule with its precious cargo arrived undamaged on the Utah desert floor on January 15, 2006. Courtesy of NASA.

Meteors

Perhaps the best way to begin our discussion of meteors, meteoroids and meteorites is to establish the nomenclature. The term "shooting stars" is the most common misnomer, which implies they are stars that somehow jump from place to place. Perhaps when you were a child you too waited in vain for a particular star to jump across the sky, confusing stars with those bright and very temporary magical displays of light that sometimes streak across the sky. Instead of "shooting" or "falling" stars, we should call them what they really are—*meteors*. People continue to use the word *meteor* to describe rocks that fall from space. But you can't hold a meteor in your hand and you can't saw it in half or hit it with a hammer. A meteor is not an object. It is a luminous effect; the light is created as a rocky body is frictionally heated to incandescence when entering a planetary atmosphere (in this case, Earth's). Often people use the term *meteorite* interchangeably with *meteor* but actually they are quite different. A meteorite is a natural rocky object of extraterrestrial origin that survives passage through Earth's atmosphere. There is one more term used throughout this book that may be somewhat confusing—*meteoroid*. The suffix "oid" is used for all rocky bodies of subplanetary size and mass in an independent orbit in space. For example, an *asteroid* is a body of subplanetary size. A meteoroid is simply a body smaller than an asteroid and probably is a chip off an asteroid "parent" body or a comet. Now let's put it all together. The *luminous* phenomena occurring when a *meteoroid* enters Earth's atmosphere is called a *meteor*. If the body survives passage through the atmosphere and lands on Earth, it is called a *meteorite*.

To be visible, the meteor must be between 100 and 150 mi of the observer, with most of that distance in altitude. Daily, over the entire Earth, the number of meteors bright enough to be visible to the unaided eye must total about 25 million. Faint meteors are far more numerous than bright ones, that is, as seen with the naked eye. If binoculars or a small telescope are used the number visible to the eye must be 100 times fainter than the eye can see alone and must approach about 8 billion daily.

The vast majority of these meteors are produced by meteoroids ranging in size from grains of sand to a few millimeters across. These do not survive their fiery passage through the atmosphere. In this size range, their brightness varies from just naked eye visibility (apparent magnitude of about +6.5) to perhaps the brightness of Venus (apparent magnitude −4.5).

Sporadic Meteors

Astronomers recognize two basic types of meteors: sporadic meteors and shower meteors. They are distinguished by their origin and predictability. Sporadic meteors are those that seem to approach Earth from any direction. Their orbits appear comet-like with no obvious preference for the ecliptic plane. Most of these meteors seem to move at speeds very close to the velocity of escape from the Solar System (46 mi/s; 74 km/s). Their orbits are nearly parabolic.

Unlike sporadic meteoroids, fireball-producing meteoroids have orbits that lie very near the ecliptic plane where the planets and most of the asteroids reside. They have small orbital eccentricities similar to main belt asteroids and their orbits are direct or counterclockwise, carrying them from west to east around the Sun. The direction at which sporadic meteoroids encounter Earth determines their atmospheric entry speed. Figure 1.9 shows how the atmospheric entry speed of a sporadic particle varies according to the direction it is traveling. Here Earth is traveling in a counterclockwise direction (west to east) around the Sun at 18.6 mi/s (29.9 km/s) as well as rotating on its axis also in a counterclockwise direction. From noon to midnight we are on Earth's trailing side facing away from its orbital direction. The vast majority of meteoroids are traveling at nearly parabolic speeds of 26 mi/s (42 km/s) as they cross Earth's orbit. To encounter Earth the meteoroids must overtake Earth from the west. The velocity of the meteoroids with respect to Earth's orbital velocity is simply the vector sum of 26 mi/s (42 km/s) minus Earth's orbital speed of 18.6 mi/s (29.9 km/s) or a minimum speed of 7.4 mi/s (12 km/s). These are *slow* meteors experienced from noon to midnight. Twelve hours later the situation reverses itself as Earth rotates on its axis. By midnight the rotating Earth has carried us into the morning sky so that we

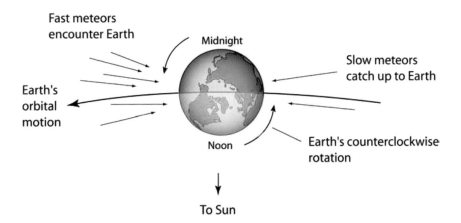

Figure 1.9. Meteoroids encounter Earth at different speeds depending on the meteor's path relative to Earth's leading and trailing sides.

now face the direction of Earth's orbital motion. The meteoroids' velocities must now be added to Earth's orbital velocity, or 44.6 mi/s (71.8 km/s). Now Earth meets the meteoroids head-on. These are *fast* meteors. The much higher encounter speeds result in faster, brighter meteors. Thus, the best time to observe sporadic meteors is between midnight and dawn.

Meteor Showers

We noted in the previous section that sporadic meteors seem to come from any direction in the sky. Their time of appearance and place on the celestial sphere is unpredictable. Meteors associated with a shower, however, appear to radiate from a small area on the celestial sphere called the *radiant*. The time and place of appearance of shower meteors are predictable. It is universally agreed by astronomers and meteoriticists that the particles come from the dust of short period comets. As these comets cross Earth's orbit they leave behind a swarm of tiny particles collectively composing a *meteoroid stream*. If the comet is relatively old, the swarm may be strewn more or less uniformly along its entire orbit so that every year Earth encounters more or less the same density of particles. This is reflected in the average number of meteors observed during the shower. A good example is the Perseid meteor shower appearing dependably every August over a four day period from August 9–12 in which about 60 meteors per hour at the peak level of activity are observed. The Perseids are among the best and most dependable of the annual meteor showers. In 1997 its parent comet, Comet Swift-Tuttle, made its perihelion passage after 105 years. Its passage rejuvenated the Perseid shower. As many as 125 meteors per hour, twice the normal count, were recorded by observers worldwide.

A comet normally does not deposit its particles uniformly along its orbit. Rather it deposits the dust in spurts. This is easily seen in the photograph made of Comet Halley during its close encounter with the Giotto spacecraft in 1986 (Figure 1.10) in which several geyser-like fountains were observed releasing their entrapped dust from sublimating ices.

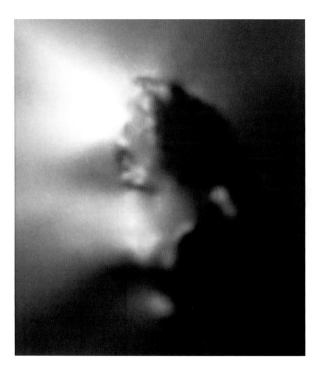

Figure 1.10. Comets do not usually deposit their dust particles uniformly along their orbits. Here, Comet Halley releases dust from two geyser-like fountains, distributing dust in spurts as it rotates. Courtesy of Dr. H. U. Keller, Max-Planck-Institut für Sonnensystemforschung, Halley Multicolor Camera Team, Giotto, ESA, copyright MPS.

Meteor showers are named after the constellation in which the radiant is found. The radiant is an effect of perspective. The meteoroids are members of a swarm traveling in closely spaced orbits parallel to each other (the meteoroid stream). As Earth passes through the swarm it is struck by many meteoroids all traveling in the same direction and with the same velocity. It is similar to viewing railroad tracks at a distance of several miles (Figure 1.11). It is an optical illusion in which the parallel-running tracks appear to converge in the distance. The swarm's meteoroids are traveling parallel to each other but in their brief sojourn through Earth's atmosphere they appear to be diverging from a radiant point in the distance. Figure 1.12 shows a photograph

Figure 1.11. A view down the center of parallel railroad tracks demonstrates how they appear to converge in the distance.

Figure 1.12. This photograph of the southern end of the constellation of Orion shows two meteors from the Leonid shower traveling parallel to each other. If they were closer to the radiant the meteors would appear to diverge away from the radiant.

Table 1.1. Eleven major meteor showers

Major meteor showers				
Shower	Maximum activity	Hourly count	Velocity (km/s)	Associated object
Quadrantids	January 3–4	100	41	Comet 141P/Machholz 2?
Lyrids	April 21–22	12	49	Comet Thatcher (1861 I)
Eta Aquarids	May 3–5	20	66	Comet Halley
Delta Aquarids	July 29–30	30	41	Comet 141P/Machholz 2?
Perseids	August 11–12	60	59	Comet Swift-Tuttle
Draconids	October 8–9	Variable	20	Comet Giacobini-Zinner
Orionids	October 20–21	25	66	Comet Halley
Taurids	November 7–8	12	28	Comet Encke
Leonids	November 16–17	Variable	71	Comet Tempel-Tuttle
Geminids	December 13–14	60	35	Asteroid Phaeton
Ursids	December 22	10	33	Comet Mechain-Tuttle

of the southern end of the constellation of Orion. Two meteors from the Leonid shower are seen traveling parallel to each other. Closer to the radiant the meteors would appear to diverge away from the radiant point. The radiant is not a geometric point. Some showers appear to diverge from a small circle only a few arc-minutes in diameter. Other radiants may be a degree or more across, twice the Moon's diameter. The radiant size depends upon how closely the particles are packed. The more densely packed the swarm, the more spectacular the shower.

Astronomy texts usually list the 10 or so most prominent and reliable annual meteor showers (Table 1.1). All of these with one single exception are produced by streams of particles from passing comets. The single exception are the Geminids occurring from December 7–15. Measurements of Geminid densities ($2.0\,g/cm^3$) compared to average comet densities ($0.3\,g/cm^3$) strongly suggests that the Geminids are the result of asteroid collisions that have liberated streams of meteoroids composed of rocky material similar to the composition of carbonaceous chondrite meteorites. In 1983 the Infrared Astronomical Satellite, IRAS, detected 3200 Phaethon, a near-Earth asteroid that was apparently sharing its orbit with the Geminids. There is one daylight meteoroid stream thought by some observers to be the debris from the asteroid 1566 Icarus. Of the 11 meteor showers listed in Table 1.1 the most prominent and trustworthy are the Perseids occurring in August, the Leonids in November, and the Geminids in December.

Early Radio Observations of Meteor Showers

In the early 1940s military radar operators noticed that meteors caused interruptions in high-frequency broadcasting reception, taking the form of whistles that rapidly dropped in pitch. Most individual meteoroids, however, are too small to reflect radar waves back to the ground. Instead, radar waves sent from the ground were detected as they reflected off of much larger targets, in this case, columns of ionized gases left in the wake of a meteor, formed when the particles evaporated passing through Earth's upper atmosphere. The first attempt to correlate visual observations of incoming meteors with radar echoes was made during the Perseid meteor shower in August 1946. At first it met with little success; the returning signals were weak at best. Then it was discovered that some radio waves, those between 4 and 5-m wavelength, are reflected efficiently only if they strike the ionized gas column of the meteor nearly broadside. Shower meteors can be detected by radar only if the radar beam is directed at right angles to the direction of the shower radiant; that is, the direction from which the meteoroids are coming.

The first detection of shower meteors was made in 1946 by Bernard Lovell and his associates at the University of Manchester using war surplus radar equipment. On the night of November 9/10, a meteor "storm" from the Giacobinids (Comet P/Giacobini-Zinner) suddenly appeared in the early morning hours and was confirmed by visual observers counting hundreds of meteors every minute. These observations confirmed the importance of radio as a means of detecting meteor activity.

Of equal importance was the discovery *of daytime* meteor showers made at Jodrell Bank in the early post WWII period. Most of these are not observable at night and only radio observations detect them during the daylight hours. In early June each year Earth passes through the densest part of two meteoroid streams, one producing between 60 and 100 meteors per hour coming from the eastern part of Aries, called *Arietids*; and the other debris stream producing up to 40 meteors per hour from a radiant in southern Perseus, called the *Zeta Perseids*. At that time hundreds of meteors streak across the sky, but unfortunately for both of these showers their radiants lie very close to the Sun in early June, making them visually unobservable. But what are serious impediments to visual observation of meteors such as moonlight, clouds and sunlight are no impediment for radio meteor detection. Interestingly, the International Meteor Organization actually lists a dozen or more daylight meteor showers that peak after sunrise and are monitored exclusively by radio means. Counting these daytime meteor showers, there are now showers that occur in all 12 months of the year. Here is a wonderful opportunity for ham radio operators to join the amateur/professional observers for the mutual benefit of both.

Meteor Outbursts and Meteor Storms

Most of the eleven meteor showers listed in Table 1.1 are not likely to produce a spectacular event. Seasoned meteor observers are happy if they can actually witness the estimated meteor rate (number of meteors per hour) associated with these annual showers. In recent years observers have applied observational corrections that result in a more realistic evaluation of the witnessed meteors. This correction value called the Zenith Hourly Rate (ZHR) standardizes the shower rate to match optimum observing conditions. Numerical values are given to varying conditions that will modify the brightness and therefore the number of meteors actually observed. By definition, the ZHR is the theoretical number of meteors which would be seen by an alert observer watching under perfectly cloudless skies, in the absence of haze and with the radiant on the zenith.

As we saw earlier, the distribution of meteoroids in a meteor stream is seldom uniform. Occasionally, Earth passes through a thick cluster of meteoroids within the main meteor stream enhancing the meteor shower for a period of several minutes to an hour or so. When a normal meteor shower is enhanced, a *meteor outburst* is said to occur. If the normal shower continues to be enhanced becoming very dense with meteoroid particles, the meteor shower becomes a full-fledged *meteor storm*. Meteor storms are generally caused by youthful meteor streams in which most of the stream's mass is still concentrated along that section of the orbit that is occupied by the parent comet. Meteor storms usually occur when Earth crosses the meteor stream at the same time as the main mass of the meteor stream is crossing the orbit of Earth. Numerically, meteor storms are defined as meteoroid counts of more than 1000 meteors per hour! In the Leonid shower of November 17, 1966, the maximum count exceeded 40 meteors per second or an unbelievable 150,000 meteors per hour and remained that high for about an hour. In all of recorded history there have been less than 10 meteor storms. Of these, seven were produced by the fabulous Leonids.

The Great Leonid Meteor Storms

Most of the time the annual Leonid meteor shower is quite poor. Literally, it's nothing to lose sleep over. Meteor observers expect the Leonids to be variable, no more than 10–15 meteors per hour. There are much better and more reliable showers from the Perseids and the Quadrantids each and every year. But the Leonids have a long and spectacular history. They are unique among the known meteor showers. It all began by accident on November 11, 1799. Two scientists, the German naturalist, Alexander von Humboldt and the French botanist Aime Bonpland, were on a five-year scientific expedition to South America when unexpectedly just before dawn on the morning of November 11 the sky lit up with thousands of meteors appearing to extend from the constellation of Leo. The two scientists learned from the local inhabitants that this was not the first time such a display had been seen. Apparently it had occurred on a more or less regular basis over the years. Humboldt's interviews with the locals suggested a roughly 30 year periodicity. This was the first time that such a periodicity (other than annually) had been noted in a meteor shower. (The *yearly* return of the Perseids was known by the early 19th century.) This investigation attracted great interest and scientists the world over waited in anticipation for the predicted return of the Leonids. On November 12, 1833, right on schedule, the night sky centered on the head of Leo provided the backdrop for one of the most spectacular meteor storms ever witnessed. Within a six-hour interval well over 200,000 meteors were recorded at several locations extending from the West Indies to Canada. By dawn the radiant point of the shower had reached the zenith. It appeared that every 33.25 years Leo would discharge thousands of meteors. After the 1833 meteor storm, astronomical historians investigating ancient records of meteor showers discovered that between AD 902 and 1833, a period of 931 years, there had been 28 noteworthy meteor showers. If these were actually 28 periodic episodes of the *same* shower recurring over and over again with an interval of 33.25 years, then the Leonids should return on November 12, 1866. And return they did, as predicted. Minor showers occurred a year or so before and after the 1866 storm as the remainder of the meteor stream passed across Earth's orbit.

After the 1866 shower astronomers began looking more closely at the radiant point and elements of the orbit of the meteor swarm. With increasingly more accurate knowledge of the position of the radiant point they began to recalculate a more accurate orbit for the swarm. They noticed that the first comet discovered in 1866 (Comet P/Tempel-Tuttle) had an orbit identical to the newly computed position of the Leonid swarm. Daniel Kirkwood, the American astronomer noted for his discovery of the gaps in the asteroid belt (Kirkwood gaps, see Chap. 2), was first to predict a connection between meteor showers and comets. Now there was no doubt. It was Comet P/Tempel-Tuttle orbiting the Sun with a period of 33.25 years that spawned billions of tiny dust particles along its orbital path, to be picked up by Earth as it crossed the dust-laden orbit of the comet. Today we know that virtually every comet is associated with meteoroid particles left in its wake.

Techniques of Observing and Photographing Meteors

Photographing meteors is not as easy as it looks. At first it seems that all you need to do is mount your camera on a tripod, aim at the radiant point, and then lie back to watch the show. The most difficult problem not easily addressed by the amateur photographer is how to point the camera in the right direction. You are trying to photograph rapidly moving objects that are not there yet. Meteor photography is the only astronomical discipline that requires lots of good fortune. First, let's look at the camera.

Film cameras are still the best and least expensive way to photograph a dark sky. Background noise in the sensors of consumer digital cameras overwhelms the small amount of information from very dim light during long exposures. Most good 35-mm film cameras come with a 50-mm focal length F/1.4 lens considered a "standard" lens in 35-mm photography. The focal length of the lens determines the angle over which the lens operates. For example, a 50-mm focal length lens covers an angle of 45° measured along the diagonal of the film frame. Shorter focal length lenses cover wider angles. A 28-mm lens, considered a standard wide angle lens covers an angle of 76°. Table 1.2 shows common focal lengths and their angles of coverage along the diagonal.

Selecting the proper focal length is essential for the meteor photographer. You must point your camera in the correct direction for the best opportunity to capture the fleeting moment when the meteor dashes across the camera's field of view. Obviously, the wider the angle covered by the lens, the more likely the meteor is to be caught passing through the field of view. Look again at Figure 1.12 which shows two Leonid meteors passing through the field of view of a 50-mm lens near the belt stars of Orion. The meteors are leaving nearly parallel trails. This parallelism occurs when meteors appear well beyond the radiant point. In this case, Orion was on the meridian in the south with the head of Leo and the radiant due east at an altitude of about 40°. Usually meteors in showers first appear about 50° in altitude and 20–30° in azimuth along their paths from the radiant. Pointing the camera toward that point will give you a better chance of capturing at least part of the meteor's path.

Another requirement for good meteor photography is a high quality fast lens; that is, a lens whose focal ratio is small. Focal ratio is simply the lens' focal length divided by its aperture, both in millimeters. For example, a lens with a 100-mm focal length and an aperture of 25 mm will have a focal ratio of 4, written F/4. This is a relatively fast lens. One word of caution is needed here. Very fast lenses tend not to produce high quality star images. Photographing a star field will help you determine the optical quality of your lens. Star images are tiny points of light that make excellent lens and telescope testers. If the star images tend to elongate toward the edge of the field of view, it is likely your lens suffers from coma. A comatic image resembles a comet with a broad sweeping tail. The "tail" points radially away from the optical center of the lens. If the star image appears to elongate in two directions at right angles to each other, the lens may be suffering from astigmatism. There are many other aberrations that may appear under star tests. Fortunately,

Table 1.2. Common focal lengths and their angles of coverage along the diagonal for 35-mm film cameras

Focal lengths (mm)	Angle of coverage (in degrees)
21	91
28	76
35	64
50	48
65	36
90	27
105	23
135	18
150	16
200	12
250	10
300	8

most of the defects can be reduced or even eliminated by stopping down the lens to a higher F/ratio. This, of course, will reduce the speed of the lens which takes us back to "square one."

Most meteor photography involves time exposures through relatively blank areas of the sky. Usually exposures of 15–20 min are made with the shutter held open with a cable release, after which the frame is advanced. Exposures beyond 20 min usually show evidence of a brightened overexposed sky due to light pollution. Using a 50-mm focal length lens the effects of Earth's rotation begins to be evident. You can expose for only about 15 s before the stars begin to leave trails produced by Earth's rotation. A 28-mm lens begins to show star trailing in about 25 s. The simplest method of photographing meteors is to place the camera on a tripod and simply expose for 15–20 min. The stars will trail from east to west at the rate of 15° per hour as Earth rotates from west to east. Some meteor photographers prefer to place their cameras on a telescope equipped with a tracking system (clock drive) to avoid star trailing. That's how the two Leonid trails below Orion were taken in Figure 1.12. Tracked images are more realistic and the faintest stars seen with the unaided eye will usually be very obvious on film. Another limiting factor is the brightness of the trail as seen in the camera's field of view. For a given focal ratio, the shorter the focal length of the lens the brighter the meteor trail will be. If you compare a photograph of a meteor trail made with a 28-mm focal length to that made with a 50-mm focal length lens, their focal ratios being the same, the shorter focal length lens will produce a brighter image even though the width and length of the meteor train will be smaller with the 50-mm lens.

As if all these details are not enough to think about when making long time exposures, you must also be aware of relative humidity. In falling temperatures the lens will often begin to dew over as the relative humidity increases. You may not be aware of this during the time exposure. It can be avoided if you use a dew shield which you can readily make for yourself. While looking through the camera, simply wrap a piece of paper around the lens barrel, extending it in front of the lens until you can just begin to see the shield as an obstruction or vignetting around the edge of the lens. This is best done by stopping down the lens to F/16 to increase the lens' depth of field. If any vignetting is occurring in the light path it will be very obvious as a dark obstructing shadow. Shorten the dew cap at half-inch increments until the shadow vanishes. This precaution seems obvious but we have heard many tales of woe from meteor photographers who simply failed to remember to include a dew shield, and subsequently lost the best meteor picture to the dewing of the lens.

References and Useful Web Sites

Books

Beech M. *Meteors and Meteorites Origins and Observations*. The Crowood Press; 2006.
Bone N. *Meteors*. Sky Publishing Corp.; 1993.
Bone N. *Observing Meteors, Comets, Supernovae and Other Transient Phenomena*. Springer Verlag; 1998.

Web Sites

The American Meteor Society and reporting a bright meteor sighting—www.amsmeteors.org
North American Meteor Network—www.namnmeteors.org
International Meteor Organization—www.imo.net
Radio Meteor Observation Bulletin—www.rmob.org
Photographing meteors www.spaceweather.com/meteors/leonids/phototips.html

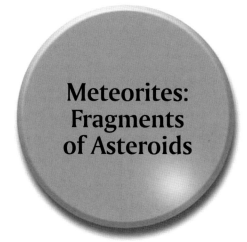

CHAPTER TWO

Meteorites: Fragments of Asteroids

What the…we've come out of hyperspace into a meteor shower. Some kind of asteroid collision. It's not on any of the charts.

Han Solo in *Star Wars: A New Hope*

Asteroids in History

In Chap. 1, we discussed the difference between meteors, meteoroids, and meteorites to be certain we are all speaking the same language. The study of asteroids also has its own evolving language. The name *asteroid* was given to these small bodies by William Herschel in 1802 in the first scientific paper on the newly discovered objects. (1 Ceres and 2 Pallas) "…resemble small stars so much as hardly to be distinguished from them. From this, their asteroidal appearance, if I may use that expression, I shall take my own name, and call them Asteroids…"

Gradually, as more asteroids were discovered astronomers began to realize that they were simply bodies of subplanetary mass and size which, in their minds' eyes gave asteroids a minor status among the Solar System's much more massive planets. By the late nineteenth century many astronomers referred to them as *planetoids* implying they were small and insignificant compared to the major planets. The final blow came when textbooks on astronomy began to refer to them as *minor planets*. After all, no telescope on Earth could resolve them into even a tiny disk. They remained star-like, even as they had been when seen by Herschel in the then largest telescope in the world (his own 48-in. reflector). Not until the 1970s did astronomers begin to realize the importance of the asteroids. These minor planets, small in stature and faint of light, would soon become giants in the struggle to understand the origin of the Solar System. A refreshing new name was finally given to them. They are *asteroid parent bodies* and their "children" are meteorites. Fragments of these parent bodies have been bombarding Earth and the other planets by the millions over the past 4.56 billion years. Hidden within these rocks from space are the clues we seek to the origin of our Solar System. It is indeed ironic that meteorites tell us more about the early Solar System than all the telescopic studies of the planets put together since the discovery of the asteroids.

Asteroids and the asteroid belt are still misunderstood by many people, particularly film makers. For example, in the movie *The Empire Strikes Back* the Millennium Falcon spacecraft encounters an "asteroid storm."

The Falcon turns into the asteroid storm and as the ship completes its turn, asteroids start coming straight at the cockpit windows. A large asteroid tumbles away from the Falcon's path at top speed. Several smaller asteroids crash into the big one, creating small explosions on its surface… The droid, Threepio, calmly calculates the possibility of successfully navigating an asteroid field as approximately three thousand, seven hundred and twenty to one.

This, of course, is fantasy. Today, thousands of asteroids are known to occupy the main asteroid belt in a zone two AU wide between the orbits of Mars and Jupiter. If you were in the main belt you would never encounter, much less have a collision with, an asteroid in your lifetime. Too much space, two few asteroids. One collision on a time scale of millions of years might be possible but never an "asteroid storm."

Main Asteroid Belt

It was Galileo's contemporary, Johannes Kepler, who first noticed the curious gap between the orbits of Mars and Jupiter. It was conspicuous because the orbits of the terrestrial planets Mercury, Venus, Earth, and Mars are remarkably symmetrical with respect to their mean distance from the Sun. As we proceed outward from the Sun, we notice that the distances between the planets increase in an orderly geometric progression. They increase in mean distance by 0.321 AU for each planet: Mercury at 0.387 AU; Venus at 0.723 AU; Earth at 1.000 AU; Mars at 1.524 AU. If we maintain this progression of distance the next planet would have a mean distance from the Sun of about 2.8 AU. No major planet occupies this position. The next planet outward from the Sun is Jupiter at 5.2 AU, clearly twice the distance if this geometric progression were to be maintained. Kepler was aware of this and was convinced that it was real. There must be an unknown planet at the 2.8 AU gap. Kepler derived an empirical relationship between a planet's mean distance from the Sun and its period of revolution: $P^2 = d^3$ where d is the planet's mean distance from the Sun in terms of Earth's mean distance in astronomical units (AU) and P is the planet's orbital period in terms of Earth's period. With this "Harmonic Law," Kepler could calculate a planet's mean distance by simply observing its period of revolution around the Sun.

The Harmonic Law along with the Law of Areas and the Ellipse Law was among the first scientific laws to emerge from the western world. Kepler himself did not know why the three laws worked. He could only trust them to what he considered to be a "divine plan" operating in the Solar System. In 1596, with only metaphysical reasons to back him, he strongly suggested that there *must* be an undiscovered planet orbiting between Mars and Jupiter. He would not live to see its discovery.

Asteroids and the Titius-Bode Rule

Over a century later, in 1766, the German astronomer Titius von Wittenburg discovered a geometrical tool that showed the spacing of the planets to be a mathematical progression. Table 2.1 shows the way it works. The progression is obtained by listing the numbers 0, 3, 6, 12, 24 48, 96, 192, 384, 768. Each of these numbers is obtained by doubling the preceding one and then adding 4 to each number. The sum is then divided by 10. The numbers obtained give the approximate distances from the Sun in terms of Earth's distance, the astronomical unit (AU).

The director of the Berlin Observatory, Johann Bode, was impressed with the rule, so much so that he used it to convince other European astronomers that there must be a planet between Mars and Jupiter. The climax came when William Herschel discovered Uranus by accident on March 13, 1781. Observations of the motions of Uranus showed conclusively that the Titius-Bode

Table 2.1. The Titius-Bode rule which describes the spacing of the planets as a mathematical progression

Titius' progression	Planet	Actual distance (AU)
(0 + 4)/10 = 0.4	Mercury	0.387
(3 + 4)/10 = 0.7	Venus	0.723
(6 + 4)/10 = 1.0	Earth	1.000
(12 + 4)/10 = 1.6	Mars	1.524
(24 + 4)/10 = 2.8	Missing planet	2.77 (1 Ceres)
(48 + 4)/10 = 5.2	Jupiter	5.203
(96 + 4)/10 = 10.0	Saturn	9.539
(192 + 4)/10 = 19.6	Uranus	19.18
(384 + 4)/10 = 38.8	Neptune	30.06
(768 + 4)/10 = 77.2	Pluto	39.4

Rule worked, at least out to Uranus, 1.5 billion mi from the Sun. The Rule showed that there should be a planet at 19.6 AU. Uranus' mean distance was measured at 19.18 AU, good enough to raise European astronomers' confidence level. It seemed they now had a rule for finding planets beyond Uranus if they existed. The next planet beyond Uranus (Neptune) according to the Titius-Bode rule should have a mean distance from the Sun of 38.8 AU. In October 1848, astronomers discovered Neptune, not by the Titius-Bode rule but by gravitational perturbations upon Uranus by an unknown planet beyond Uranus. The Titius-Bode rule fails completely for Neptune and Pluto.

Discovery of the First Asteroids

After Uranus was discovered other astronomers, their interest renewed, took up the search for the elusive planet that had to exist between Mars and Jupiter. In Palermo, on the island of Sicily, where the most southerly European Observatory had been completed a decade earlier, Giuseppe Piazzi had been working on a new star chart which could be used to search for the presumed planet. On the night of December 31, 1800, he spotted an 8th magnitude star near the ecliptic in the constellation Taurus. This star was not on the charts he had been revising. He positioned the star on the new chart and awaited the following night. The next night the star had moved again relative to the stationary background stars. Over the next several weeks, he plotted the motion of the star through the constellation Taurus. On February 11, 1801, Piazzi became ill and was forced to end the observations. Fearful of losing the object, he contacted Bode at the Berlin Observatory and reluctantly made his observations known. Understandably, Piazzi wanted to keep the observations a secret until he had an opportunity to plot the section of the orbit he had observed. It wasn't until late spring that he finally disclosed the positional data he had acquired, first to Bode in Berlin (May 31) and then to J. J. Lalande at the Paris Observatory (June 11). In this interim and without confirmation from the other observatories, Piazzi named the new "planet" 1 Ceres, the Roman Goddess of agriculture. By this time, the new "planet" had changed its position and it could not be located at either of these observatories. No matter how diligently these European astronomers searched the skies, the new "planet" 1 Ceres was promptly lost.

But the story of the first asteroid to be discovered does not end there. Several observatories searched for it with no luck. Then Europe's most distinguished mathematician, Karl Gauss, joined the search. Gauss was able to calculate an orbit from the original data Piazzi had acquired up to February 11, 1801. From this data, he pinpointed its location. On January 1, 1802, exactly 1 year after of the initial discovery, 1 Ceres was rediscovered and never lost again. (Figure 2.1)

Figure 2.1. False color image of 1 Ceres taken on January 24, 2004, by the Hubble Space Telescope. Courtesy of NASA, ESA, J. Parker (Southwest Research Institute), P. Thomas (Cornell University), L. McFadden (University of Maryland), and M. Mutchler and Z. Levay (STScl).

The recovery of 1 Ceres was the beginning of a series of discoveries made over the next 7 years that culminated in the discovery of three additional asteroids: 2 Pallas, 3 Juno, and 4 Vesta. It became apparent that the space between Mars and Jupiter was not the home of a single planet but of several "minor planets," each with its own orbital characteristics which placed them in a zone called the *asteroid belt*. The vast majority of asteroids are found in the main belt between 2 AU and 4 AU from the Sun. The discovery of new asteroids after the first four had to wait 38 more years. Finally, the wait was over when, in 1845, a German amateur astronomer announced the discovery of the fifth asteroid, Astraea. The discovery prompted a renewal of the asteroid race but by this time (early nineteenth century) a new technology had appeared that was destined to change everything—astronomical photography. Before the invention of photography, both amateur and professional astronomers had only their eyes glued to the eyepiece of a telescope to help them make their discoveries. But this dependency on the human eye at the telescope was short-lived. By the mid-nineteenth century, dry plate photography was rapidly taking over and long time exposures on film were capable of picking up asteroid images hundreds of times fainter than the human eye alone could detect. By the close of the nineteenth century over 300 asteroids were known. By the mid-twentieth century, over 4,000 asteroids had been located and their orbits determined. There seemed to be no end to these little worlds. The minor planets ruled supreme. Now, there are over 30,000 known.

The last quarter of the twentieth century saw the most extraordinary advance in astronomical imaging in the history of observational astronomy. The development of charge coupled devices (CCDs) with sensitivities hundreds of times that of the fastest films rapidly engulfed the study of asteroids. Asteroid astronomy has become digital. Now, automated electronic telescopes nightly comb the skies for these chunks of rock. And amateur astronomers are not far behind. Today, they have equipped themselves with large aperture commercially made telescopes with sensitive CCD electronics that only professional observatories had possessed just a decade or two earlier. With relatively large amateur telescopes equipped with digital electronics, images of asteroids to the 16th magnitude or fainter have become possible. Amateurs have significantly added to the discovery rate. If this continues, in only 3 or 4 years there will be a doubling of known asteroids.

Cataloging and Naming New Asteroids

After the end of World War II, the International Astronomical Union (IAU) established the Minor Planet Center where observational data from amateur and professional astronomers world-wide could be sent for analysis. Each month huge volumes of data pour into this data bank located at the Smithsonian Astrophysical Observatory in Cambridge, Massachusetts. Here, data on new discoveries of comets and asteroids are processed. The first step is to feed the data from the new discovery into computers to see if there is a match with known or suspected comets or asteroids. If the object appears to be a new asteroid, a temporary provisional designation is given the object. The object must have been observed over at least two consecutive nights to be eligible for this temporary designation. The provisional number is a combination of the year and month of discovery. The position of the provisional asteroid must now be compared with known asteroid positions (or other provisional asteroids) through at least one opposition of the object in question. If no link can be made with other provisional asteroids, then further observations are made, from which an orbit is computed. At this time, the asteroid could be followed through several more months to refine the orbit and search for other links with provisional asteroids. Finding no further links at this time, it is highly probable the object is indeed a new asteroid. Once astronomers have reached this certainty, the asteroid is given its final designation—a number preceding its name that denotes the numerical order of discovery and finally a name. Thus, 1 Ceres, the first asteroid discovered would be given the number 1 and the name, 1 Ceres. Other asteroids follow 1 Ceres in numerical order. Naming the newly discovered asteroid is a much simpler process. The finder has the privilege of providing the name. When asteroids were first found they were given female names from classical Greek and Roman mythology. At the time, astronomers had no idea that these names would quickly be depleted as the number of asteroid discoveries climbed rapidly through the nineteenth century. It is no surprise that this privilege soon led to some rather inappropriate names—from rock stars to historical characters of rather dubious distinction. Finally, in 1982, the IAU formed the Small Bodies Names Committee whose purpose was to examine each name and judge its suitability for publication.

From Asteroid Belt to Earth

It is generally accepted among planetary astronomers that asteroids are parent bodies of the meteorites. That meteorites are fragments of asteroids has yet to be proven but is provisionally accepted as a working hypothesis. What is questioned is how meteorites manage to make it to Earth. There are three zones in the Solar System where asteroids are known to reside. The best known is the *main belt* which occupies a zone between 2 AU and 4 AU from the Sun. It is a zone of considerable stability. Most of the main belt asteroids have been there in near circular orbits about the Sun since their formation 4.56 billion years ago. By 1866, a sufficient number of asteroids had been discovered in this region to establish the reality of the zone, but something was amiss. In that year, the American astronomer Daniel Kirkwood pointed out that the Belt was not as uniform as first thought. There seemed to be gaps in the belt in which few or no asteroids were found. Figure 2.2 shows the distribution of asteroids within the belt. The gaps are obvious, but what was causing them? Kirkwood and other astronomers knew that Jupiter's gravitational force was probably responsible for initially herding the asteroids into the main belt. But more importantly, he recognized that some of the asteroids near or within the gaps had unstable orbits. Kirkwood calculated the orbital period of the gap positions and found that the periods were related to Jupiter's orbital period, that is, the asteroids in the gap positions have orbital periods that were simple fractions of Jupiter's period. For example, an asteroid at 3.28 AU has a period of 5.94 years which is exactly half of Jupiter's orbital period of 11.88 years. Thus, every 2 years Jupiter

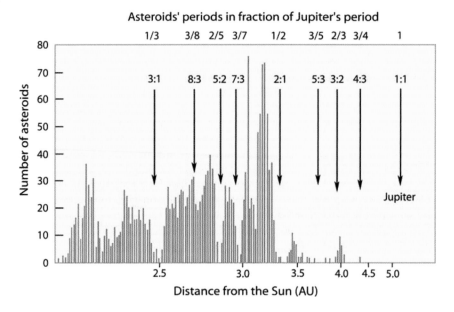

Figure 2.2. A plot of asteroid distance versus orbital period shows obvious gaps in the asteroid belt produced by Jupiter's gravitational perturbations. These are known as the Kirkwood gaps.

and that asteroid will experience a close encounter. This gravitational link is called *resonance*. Any asteroid whose period is a simple fraction of Jupiter's period will experience gravitational perturbations much more often than the other asteroids in stable orbits. The result of these perturbations is that within a few million years the orbital eccentricities of these perturbed asteroids will gradually increase, accelerating them along elliptical orbits and carrying them across the belt, raising the probability of collision with other main belt asteroids. Such collisions may be the mechanism that sends fragments across the asteroid belt toward Earth and the terrestrial planets. If they survive passage through the main belt they may eventually establish elliptical orbits among the inner planets.

Near-Earth Objects

Near-Earth Objects or NEOs are those asteroids that have escaped the confines of the main belt. They roam freely among the planets of the inner Solar System in which Earth is the largest target. Asteroid 433 Eros was the first discovered to have left the main belt and crossed the orbit of Mars. Eros comes within 13 million miles of Earth's orbit. In March 1932, another Mars-crossing asteroid was discovered with a perihelion of 1.08 AU. It was given the name *1221 Amor*. It became the prototype Mars-crosser with a perihelion between 1.0 AU and 1.3 AU. About a month later, another near-Earth asteroid was found but this time with a perihelion inside Earth's orbit. This asteroid was moving very swiftly when discovered, indicating its nearness to Earth. 1862 Apollo was designated the prototype Apollo asteroid having perihelia well inside Earth's orbit. Further observations showed that Apollo actually passed inside the orbit of Venus. Collectively, they were given the name *Earth-crossing* asteroids. Both Amors and Apollos still maintain their ties with the main belt, having their aphelia within the confines of the main belt. Inevitably, in 1975, another asteroid was discovered that had entirely broken its ties with the asteroid belt. Its mean distance lay entirely within the Earth's orbit. This asteroid, 2062 Aten, became the prototype for *Aten* asteroids.

Trojan Asteroids

There is one last asteroid group that we should include in this brief survey. The very dark bodies in this group share a mutual orbit with Jupiter. There are actually two separate groups found 60 degrees east and west of Jupiter (Figure 2.3). Discovery of the first *Trojan* asteroid was made by the German astronomer Maximilian Wolf in 1906. It was not a complete surprise that such asteroid clusters existed. The French mathematician Joseph Lagrange in 1772 showed that as many as five such clusters could exist "attached" to Jupiter's orbit. However, of the five *Lagrangian* points, only two (L_4 and L_5) are stable. The Trojan asteroids are at the 1:1 resonance point where small asteroids can remain stable for an extended period. The largest Trojan asteroid is 624 Hector which measures ~190 by 95 mi (300 by 150 km). The total population, averaging 9 mi (14 km) or less in diameter, may number in the thousands and are comparable to the smaller asteroids in the main belt.

An Important Job for Dedicated Amateur Astronomers

Today we know of 800 or more Near-Earth asteroids. There must be many others awaiting discovery. Here is a wonderful research opportunity in which amateur astronomers can become involved and make an important contribution. The situation is that there are too many asteroids

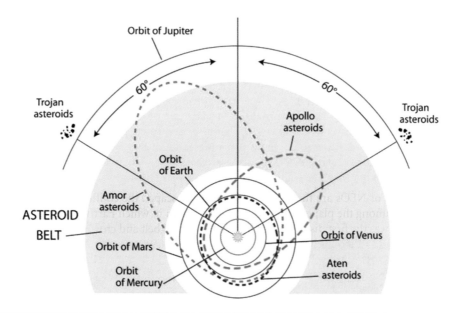

Figure 2.3. Typical orbits of Apollo, Amor, and Aten asteroids, all of which cross Earth's orbit. They are collectively termed Near-Earth asteroids. Trojan asteroids occupy very stable positions 60 degrees east and west of Jupiter's position.

and not enough observers. We saw that there are currently over 30,000 asteroids known, the vast majority from the main belt. In addition, there may be as many as 1,000 or more NEOs. All of these asteroids are affected by gravitational perturbations from Jupiter and Saturn that, in only a few years, will have changed their orbits sufficiently so that they become lost. (Remember the first asteroid to be discovered, 1 Ceres, was lost within months of its discovery, not to be found again for another year.)

Astronomers involved in the surveys (five currently operating—see below) report their NEO discoveries to the Smithsonian Astrophysical Observatory's Minor Planet Center located at Harvard University, Cambridge, Massachusetts. These observations are posted to a Website called the NEO Confirmation Page. Once an object is designated an NEO, it must be continually tracked to assure that it not be lost. There is always some uncertainty in their orbits that only increase in time. They need to be observed on a regular basis. There are too many newly discovered NEOs that require follow-up observations. Here is where the dedicated amateur astronomer can help. Amateur astronomers play a critical role by relocating and further tracking each new NEO discovery. These observations help refine the object's orbital data and assure that the asteroid will not be lost in the future.

The Five Major NEO Surveys

MIT Lincoln Lab and Air Force LINEAR project in Socorro, New Mexico

University of Arizona Spacewatch Survey, Steward Observatory, Tucson, Arizona

Lowell Observatory LONEOS at Flagstaff, Arizona

Jet Propulsion Laboratory, NASA and Air Force NEAT program in Hawaii

Catalina Sky Survey in Tucson, Arizona

Comparing Asteroids with Meteorites

Virtually all astronomers believe that meteorites are pieces of asteroids although they have never sampled an asteroid directly. We know that meteorite collections world-wide must contain pieces of at least 135 separate asteroids. These probably do not include all of the asteroid types. About 85% of all meteorites that have made it to Earth are ordinary chondrites. This tells us that the meteorites in our collections are probably biased toward the ordinary chondrites. What we need is to find a way to compare the mineralogy of meteorites in the laboratory to the mineralogy of asteroid parent bodies in the asteroid belt. Simply said, but not so simply done; however, there is a way. The Galileo spacecraft had near encounters with the asteroids 951 Gaspra and 243 Ida, both probably related to ordinary chondrites. Their surfaces appeared to be covered with blankets of loose rocky material that rested on a consolidated layer of bedrock. This material accumulated after countless millions of years of repeated impacts by small meteoroids. This surface is called a *regolith*. It is a major surface covering much of the Moon as well as the surfaces of main belt asteroids. Much of the material has experienced a history of fragmentation and compaction during countless episodes of cratering. During collisions, angular rocks were broken into smaller rocks and then cemented into a hard rocky material called a regolith *breccia*. Some regolith breccias from carbonaceous chondrites and ordinary chondrite meteorites have actually made it to Earth intact.

The surface regolith material on the smallest scale is a mixture of fragmented and compressed rock made of tiny mineral grains. Sunlight reaching the surface of an asteroid is either absorbed, transmitted through or reflected by these grains. The ratio of incident light to reflected light is called the *albedo* which is defined as the reflective efficiency of the tiny grains. The reflectivity depends upon how each mineral responds to the visible and infrared spectrum. As sunlight passes through the mineral grains, the grains absorb specific wavelengths and reflect back a solar spectrum minus the absorbed wavelengths. The mineral crystals on the surface do not produce sharply defined absorption lines like we see on the Sun; rather, we see a composite of broad dark bands composed of several minerals making up the surface spectra. These are sorted out by comparing the spectra with laboratory reflectance spectra of purified minerals.

Only a few minerals produce prominent infrared absorption features measured by a reflectance spectrophotometer. Fortunately, these are the very minerals that are found in chondritic meteorites. Figure 2.4 shows reflectance spectra of the three primary chondritic minerals: pyroxene, olivine, and iron (metal). Using reflectance spectra, a classification of asteroids was worked out. Table 2.2 summarizes the important asteroid compositional types, relating them to their albedos, meteorite association, and their approximate location in the asteroid belt. The table is arranged in order of decreasing albedo but it is also arranged in order of increasing distance from the Sun. For example, type E asteroids are close to the inner belt and also to the Sun. They are chemically related to meteorites called *aubrites*. Type V is related to the asteroid 4 Vesta. It is a differentiated asteroid that has undergone numerous collisions in its history. We will return to 4 Vesta shortly.

Type S asteroids are believed to be related to *ordinary chondrites*, the most commonly found meteorites on Earth. They are located between the middle and inner asteroid belt. Type M asteroids are found in the central belt and may be related to the iron-bearing E chondrites and the iron meteorites. Type D are Jupiter's Trojan asteroids found on the L_4 and L_5 Lagrangian points some 60° on either side of Jupiter and beyond the extreme outer edge of the main belt. They are very dark and metallic. Type C carbonaceous asteroids are most abundantly found in the main belt and may be related to the CM2 carbonaceous chondrite meteorites.

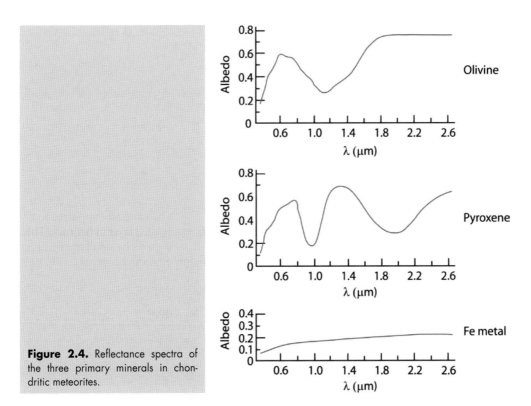

Figure 2.4. Reflectance spectra of the three primary minerals in chondritic meteorites.

After asteroid reflectance spectra are made, the next step is to select meteorites in the laboratory that have similar reflectance spectra for comparison. Most asteroids are covered with a regolith of broken fragments with fine-grained minerals cementing them. To make asteroid/meteorite comparisons the surface characteristics of both must closely match. This simulation is best done by grinding the meteorite to a fine crystalline powder to make the optical qualities as similar as possible. Figure 2.5 shows a few of these comparisons. Here, reflectance is plotted against the visible and infrared wavelength. The lines show laboratory reflectance spectra of five common

Table 2.2. Classification of asteroids arranged in order of decreasing albedo

Type	Albedo (%)	Associated meteorite	Location
E	25–60	Aubrites	Inner belt
A	13–40	Pallasites, olivine-rich	Main belt (?)
V	40	Eucrites, basaltic	Middle main belt, 4 Vesta and fragments
S	10–23	OC (?), mesosiderites	Middle to inner belt
Q/R	Like S	Possibly unweathered OC, with variable olivine/pyroxene	Middle to inner belt
M	7–20	E-chondrites, irons	Central belt
P	2–7	Like M, lower albedo	Outer belt
D	2–5	Trojans	Extreme outer belt, Jupiter L_4, L_5 points
C	3–7	CM carbonaceous chondrites	Middle belt, 3.0 AU
B/F/G	4–9	C subtypes	Inner to outer belt

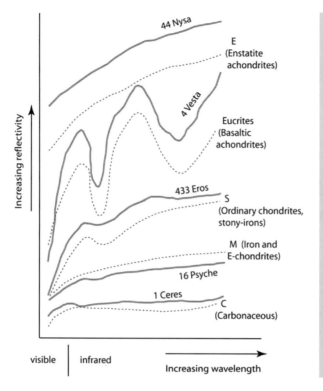

Figure 2.5. A comparison of a few selected asteroid and meteorite reflectance spectra from the visible spectrum to the infrared and differing in mineralogy. These are compared to asteroid reflectance spectra of surface minerals made with telescopes from Earth.

meteorites. The solid lines show asteroid reflectance spectra. The comparison shows a close match between the asteroid surfaces and the mineralogy of the powdered meteorites. In particular, the spectrum of 433 Eros closely matched an L4 ordinary chondrite, as did the spectrum of the Apollo asteroid 1685 Toro (not shown).

C-type carbonaceous asteroids are the most abundant of the asteroids found in the main belt. They are dark bodies with albedos between 3% and 7%, only half the albedo of the Moon. More than half of the C-type asteroids show signs of combined water.

The S-type asteroids are the second largest group in the main belt. They probably represent the closest match to the ordinary chondrites. Here a conundrum arises. On Earth the ordinary chondrites outnumber all other meteorite types by a large margin. Yet only about 16% of the S-type asteroids studied have chondritic compositions. The disparity between the carbonaceous chondrites, rare on Earth but plentiful in the asteroid belt, and the ordinary chondrites, common on Earth but relatively rare in the asteroid belt, seems to be telling us that ordinary chondrites probably came from one or at most a few asteroid parent bodies. The large numbers of chondrites reaching Earth are apparently not indicative of large numbers in space. Thus, it does not seem that we can rely upon our meteorite collections to tell us the true ratios of asteroid types.

4 Vesta

In 1970, T.B. McCord and his coworkers at the Institute of Geophysics and Planetology, University of Hawaii, made astronomical history when they were the first to recognize similar characteristics between the spectra of 4 Vesta and a specific meteorite type. They compared the reflection spectra of the Nuevo Laredo achondrite with the reflection spectra of 4 Vesta. Nuevo Laredo is a member of the HED clan of achondrites, specifically, a eucrite. It was the first successful effort to relate an asteroid to a specific meteorite type.

4 Vesta was a good choice for the analysis in that it is one of the largest asteroids known at 330 mi (530 km) in diameter and it is occasionally bright enough to be seen with the unaided eye. Amateur telescopes easily show it as a 5th magnitude star-like object moving slowly among the stars. 4 Vesta rotates on its axis with a period of 5.3 h. As it rotates, its spectrum constantly changes. This can only mean that it is not a homogeneous body, but a heterogeneous one. Its surface composition constantly changes in time as it rotates. In some areas, the asteroid shows a surface that is primarily eucritic, meaning that it is basaltic in composition. The eucritic crust is interpreted as areas where lava flows had erupted from beneath the surface and spread over the landscape. Some of the surface has been impacted by other asteroids forming impact craters. These craters pass through the thin eucritic crust to an intrusive layer in 4 Vesta's upper mantle. This is composed of a plutonic layer with a diogenitic composition. (HED stands for Howardite–Eucrite–Diogenite, related basaltic meteorites believed to originate on 4 Vesta.) The deepest layers exposed on 4 Vesta may be composed of olivine-rich material (Figure 2.6).

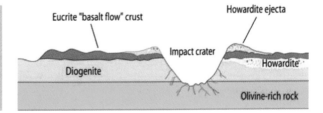

Figure 2.6. Diagram of asteroid 4 Vesta showing surface structure and interior cross section of the crust and upper mantle.

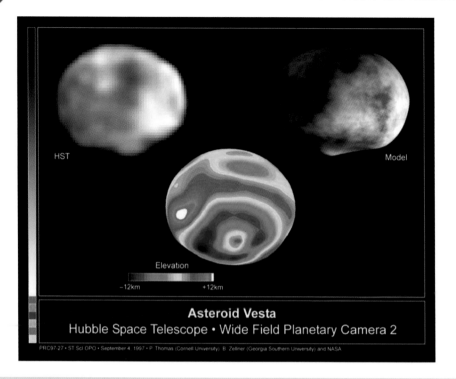

HST

Model

Elevation

−12km   +12km

Asteroid Vesta
Hubble Space Telescope • Wide Field Planetary Camera 2

PRC97-27 • ST ScI OPO • September 4, 1997 • P. Thomas (Cornell University) B. Zellner (Georgia Southern University) and NASA

Figure 2.7. Hubble Space Telescope images of 4 Vesta revealing surface features including a huge impact basin 456 km in diameter and nearly 13 km deep. Photo taken September 4, 1997. Courtesy NASA, Ben Zellner (Georgia Southern University), and Peter Thomas (Cornell University.)

In 1997, the Hubble Space Telescope photographed a series of images that revealed a huge impact basin near Vesta's south pole measuring 283 mi (456 km) wide that covered more than 75% of one side (Figure 2.7). The impactor was about 18 mi (30 km) in diameter and struck Vesta at a speed of 3.0 mi/s (4.8 km/s) leaving behind a crater 7.9 mi (12.8 km) deep. Although huge chunks of Vesta were undoubtedly expelled, the impact was not energetic or massive enough to disrupt Vesta but countless meteorite-sized fragments must have been blasted into space. It was estimated that about 1% of the material that makes up Vesta must have been excavated at the time of impact. Further Hubble images showed a knob-like structure near the South Pole of the asteroid which has been interpreted as an impact basin almost as large as Vesta itself! As the resolution increased in time, a central uplift or rebound peak was revealed, similar to rebound peaks centrally located inside impact craters on the Moon. It is probable that the formation of this impact basin is responsible for the origin of the HED achondrite meteorites.

1 Ceres

Earlier we reviewed the story of the discovery of 1 Ceres, the largest asteroid in the main belt. 1 Ceres is a member of the C-type asteroids (as is Pallas, the second asteroid discovered). It is a nearly spherical body about 584 mi (940 km) in diameter and is among the most primitive asteroids known. It is very dark, having an albedo of only about 5%. A broad absorption band in its spectrum at 3.0 μm indicates the presence of water-bearing clay minerals or phyllosilicates. 1 Ceres is most closely compared to members of the carbonaceous chondrite groups, especially the CI and CM chondrites that have suffered severe aqueous alteration.

In 2006, the International Astronomical Union (IAU) reclassified 1 Ceres as a dwarf planet. So the largest asteroid is now the smallest dwarf planet, one of three. The others are Eris and Pluto.

Asteroid Close Encounters

We have seen how difficult it is to observe main belt asteroids even with the largest Earth-based telescopes. And even under the best of conditions the Hubble Space Telescope could only image the largest asteroids as small, nearly featureless disks. But that was about to change. On October 29, 1991, the Galileo spacecraft making its way to Jupiter encountered the S-type main belt asteroid 951 Gaspra. Images were obtained from a distance of 3,190 mi (5,150 km). It was immediately obvious that Gaspra was a relatively small irregular fragment splintered off a much larger asteroid (Figure 2.8). It measured 11.3 × 5.5 × 6.5 mi (18.2 × 8.9 × 10.5 km). Gaspra showed a surprising lack of large impact craters, most appearing almost erased from the landscape. This sheared-off fragment apparently exposed a fresh surface with few if any large impact craters. Two years later while still on its journey to Jupiter, the Galileo spacecraft again made history when it encountered a much larger asteroid. It passed by 243 Ida from a distance of only 6,790 mi (10,950 km) (Figure 2.9). This was another S-type asteroid. In contrast to Gaspra, Ida was heavily

Figure 2.8. Galileo spacecraft image of asteroid 951 Gaspra, the first asteroid image made from space. Galileo spacecraft image made October 29, 1991. Courtesy of NASA.

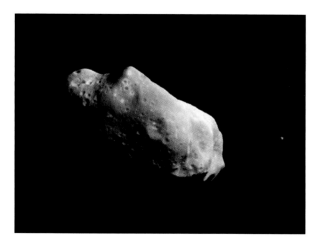

Figure 2.9. Galileo spacecraft image of asteroid 243 Ida taken August 28, 1993, from a distance of 10,950 km. Ida is an S-type weathered asteroid. Newly discovered satellite, Dactyl, on far right is only 1.5 km in diameter and 48 km from Ida's surface. Courtesy of NASA.

cratered. It was also a fragment of a much larger parent body, and measured 37.1 mi (59.8 km) in length. During Ida's flyby, an astonishing discovery was made. A small moon 1 mi (1.5 km) in diameter was orbiting Ida at a distance of only 30 mi (48 km) from the surface. This little satellite, now called Dactyl, was the first moon discovered orbiting around an asteroid.

Both Gaspra and Ida with its moon show S-type spectral characteristics different from the ordinary chondrites measured in the laboratory. Studying the surface spectral characteristics strongly suggest that the optical characteristics of the surface minerals of S-type asteroids have been altered due to *space weathering*.

253 Mathilde

On February 17, 1996, asteroids once again made the news when the NEAR (Near Earth Asteroid Rendezvous) spacecraft was launched. The first two asteroid encounters were brief opportunities which took a back seat to the primary Galileo/Jupiter mission. The primary mission was to be the near-Earth asteroid 433 Eros. But this asteroid mission had a secondary target, the asteroid 253 Mathilde. On June 27, 1997, the NEAR spacecraft passed to within 751 mi (1,212 km) of Mathilde. This asteroid was discovered over a century ago, but it wasn't until 1995 that ground-based observations showed Mathilde to be a C-type asteroid with an albedo of only 4%, the brightness of charcoal, half the albedo of the dark mare on the Moon. This low albedo strongly suggested that Mathilde was a fragment of a CM carbonaceous chondrite. Further density studies gave Mathilde a bulk density of only 1.3 g/cm³, half the bulk density of a typical CM chondrite. This could only mean that Mathilde has a rubble pile internal structure.

There are two models that could describe the interior of a chondritic asteroid parent body. The original body is accreted as it orbits in the protoplanetary disk. The result is a homogeneous body with its mineral components evenly distributed throughout the interior. Internal heating by the short-lived radioisotope ^{26}Al provides the energy to heat the interior from the deep core of the body to the near surface. Thermal metamorphism slowly heats the interior to a petrographic type 6 at the core. The heat makes its way through the body, slowly converting various regions of the interior to different petrographic types from type 6 to type 3. The result is a layered structure something like an onion's interior, thus, the *onion shell model*. Sometime early in its history after the onion shell had formed, the asteroid parent body was catastrophically disrupted by impact with another asteroid. The impact broke the body into myriads of smaller fragments, but in this case the impact was not strong enough to scatter the pieces. Instead, mutual gravity between the pieces reassembled the rubble pile into a mixture of petrographic types (Figure 2.10).

Today, Mathilde is heavily scarred with two enormous impact craters, one in deep shadow (Figure 2.11). The other sits on top of the asteroid when viewed nearly edge on. The surface is smooth and remarkably uniform, suggesting a homogeneity that probably runs through the entire body. Mathilde is a very ancient, primitive world.

433 Eros

Unlike the asteroids explored above, 433 Eros does not reside in the main asteroid belt. It is a Near Earth Asteroid that has its aphelion well inside the asteroid belt but its perihelion just inside Earth's orbit. Figure 2.12 shows the path taken by the NEAR spacecraft on its extraordinary mission to Eros. The NEAR EARTH ASTEROID RENDEZVOUS mission was launched on February 17, 1996, and after a brief encounter with Mathilde received a gravity assist as it encountered Earth on January 23, 1998. NEAR was scheduled to arrive at Eros on December 23, 1998, but within 3 days of the arrival a malfunction of its main engines caused orbit insertion to be aborted. The burn had

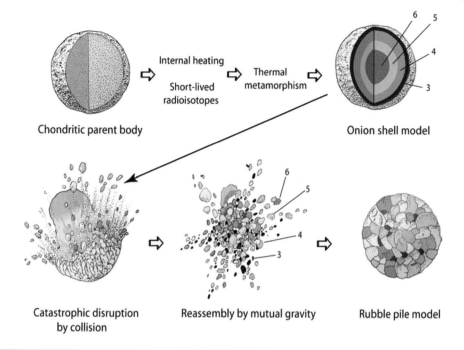

Chondritic parent body

Internal heating

Short-lived radioisotopes

Thermal metamorphism

Onion shell model

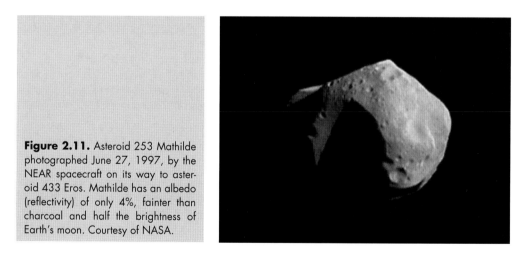

Catastrophic disruption by collision

Reassembly by mutual gravity

Rubble pile model

Figure 2.10. Onion shell versus rubble pile models of a chondritic asteroid parent body.

Figure 2.11. Asteroid 253 Mathilde photographed June 27, 1997, by the NEAR spacecraft on its way to asteroid 433 Eros. Mathilde has an albedo (reflectivity) of only 4%, fainter than charcoal and half the brightness of Earth's moon. Courtesy of NASA.

stopped after less than 2 seconds. Mission operators had lost contact with the spacecraft. NEAR was tumbling in space with no guidance. After a grueling period of 27 hours contact was finally reestablished but the opportunity to rendezvous and orbit around Eros on December 23, 1998, as planned, was no longer possible. If nothing else happened to control and redirect the spacecraft, NEAR would coast by Eros on the 23rd of December at a distance of 2,378 mi (3,827 km). The spectacular saga of NEAR's eventual rendezvous and orbit around Eros beginning on Valentine's Day, February 14, 2000, is an extraordinary story. It is dramatically presented in the book Asteroid Rendezvous—NEAR Shoemaker's Adventures at Eros, published by Cambridge University Press, 2002.

Near Earth Asteroid Rendezvous

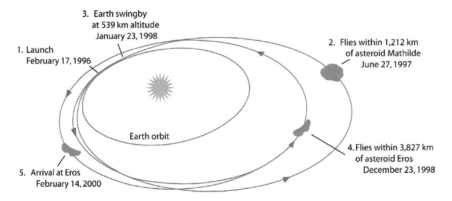

Figure 2.12. Trajectory of the NEAR spacecraft to 433 Eros. Courtesy of the Applied Physics Laboratory, Johns Hopkins University, and NASA.

The year 2000 will be remembered by planetary astronomers and meteoriticists throughout the world as the year NEAR orbited 433 Eros. This was the first opportunity to study a near-Earth S-type asteroid for a full year. Among the many instruments carried by the NEAR spacecraft was a near-infrared spectrometer (NIS) designed to scan the surface and measure the spectrum of sunlight reflecting off Eros at wavelengths between 0.8 and 2.7 μm. The goal was to determine the composition, distribution and abundance of surface minerals. For 3 months after initial orbit insertion, the NIS instrument took over 200,000 reflectance spectra of Eros'surface. The spectra confirmed the presence of olivine and pyroxene on the surface. Moreover, the ratio of olivine to pyroxene was similar to that found in ordinary chondrite meteorites. The stage was set.

On May 13, 2000, the NIS instrument unexpectedly failed, but fortunately not before 90% of the surface had been scanned. NEAR had provided strong evidence that Eros was indeed a fragment of a larger asteroid parent body, and that body was an L4 ordinary chondrite.

Hayabusa

Formerly called MUSES-C, the Hayabusa (the name means falcon) mission is in many ways similar to the Eros mission except that it was designed to bring back samples from a small near-Earth asteroid named 25143 Itokawa. The Hayabusa spacecraft employs new technology designed to use, for the first time in space, an ion propulsion engine. This engine ionizes and accelerates the propellant gas, xenon, and then electrically emits these accelerated particles into space.

Until now, only the extraterrestrial material gathered during the Apollo Moon missions has been used for Earth-side analysis. Overall chemical compositions of the Moon and planets can be determined from Earth-based telescopes and compared with meteorite samples scattered across the Earth's surface. But the Moon and planets have evolved over time, changing due to thermal processing of the Solar System's earliest parent bodies. Thus, the Moon and other planetary bodies cannot provide us with a pristine record of the early Solar System. Asteroids, however, are thought to be small enough to have preserved the physical state of the early Solar System. Soil and rock samples from a small asteroid like Itokawa can give us important clues about the raw materials that made up asteroid parent bodies in their formative years 4.56 billion years ago.

Hayabusa was launched on May 9, 2003, and rendezvoused with Itokawa in mid-September 2005. Hayabusa first surveyed the asteroid's surface from a distance of about 12 mi (20 km) and then moved in for a much closer look. The spacecraft studied Itokawa's numerous physical parameters such as its composition, color, density, shape, and topography. Then, on November 4, 2005, Hayabusa attempted to land on Itokawa but was unable to complete the maneuver. At the second landing, the spacecraft was programmed to fire tiny projectiles at the surface and then collect the resulting spray in its "collection horn." The second attempt failed, but on November 19 Hayabusa set down on the surface. It lifted off without taking a sample; however, there is a high probability that on a later attempt some of the dust from the surface made its way into Hayabusa's sampling chamber which is now sealed for the journey home.

Hayabusa was not specifically designed to land on Itokawa but only to touch the asteroid's surface with its sampling tool. Though not part of the original mission plan, Hayabusa was indeed able to land on the asteroid's surface, remaining there for about 30 minutes. before resuming flight. This is the first time a spacecraft has landed and taken off from a solar system body other than the Moon (Figures. 2.13–2.15).

Itokawa has been shown to be a chondritic rubble pile that has never undergone differentiation into core and mantle. Instead, for millions of years, it has suffered countless impacts from the rocks still scattered everywhere on its surface. Amazingly, Itokawa has no impact craters, only a rocky boulder-ridden surface. Near the "Muses Sea" where smooth terrain interrupts the otherwise rocky surface, the first successful historic touchdowns were made.

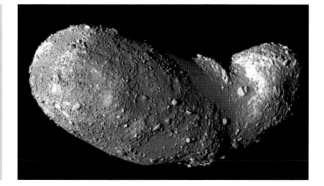

Figure 2.13. The asteroid Itokawa photographed by the Hayabusa spacecraft on October 23, 2005, from a distance of 4.9 km. Some areas on the surface are very smooth and ill-defined, notably the area known as the Muses Sea near the center of the asteroid. Researchers believe this is the result of vibrations caused by impacts on a "rubble-pile" (loosely consolidated) asteroid. Copyright ISAS/JAXA.

Figure 2.14. Itokawa seen from its north pole on November 1, 2005. The longest axis is 535 m and the other axes are 294 and 209 m. Itokawa appears to be composed of two or three individual segments reassembled after multiple impacts. Copyright ISAS/JAXA.

Figure 2.15. Rocks broken by countless impacts lie scattered on the surface of Itokawa, seen from a distance of 0.11 km. Taken on November 12, 2005. Copyright ISAS/JAXA.

The Dawn Mission to 4 Vesta and 1 Ceres

The Dawn mission to 4 Vesta and 1 Ceres was cancelled due to cost overruns and technical problems. But the human drive to understand our origins is a drive that cannot easily be controlled, much less denied. On March 26, 2006, NASA senior management officials announced to the world that the Dawn Mission had been reinstated. "We revisited a number of technical and financial challenges and the work being done to address them. Our review determined the project team has made substantive progress on many of this mission's technical issues. And in the end, we have confidence the mission will succeed." (NASA Associate Administrator Rex Geveden, chairman of the review panel.)

The mission was named because it was designed to explore and study objects dating from the dawn of the Solar System 4.56 billion years ago. The mission's objectives are to send a spacecraft to asteroids 4 Vesta and 1 Ceres, among the largest asteroids orbiting the Sun between Mars and Jupiter. Of the thousands of asteroids orbiting in the main belt, asteroid 1 Ceres is the largest at 584 mi (940 km) in diameter, while 4 Vesta ranks 4th at 330 mi (530 km) in diameter. The two asteroids are quite different from each other. 1 Ceres is the most primitive. It contains both water and organic compounds much like carbonaceous chondrite meteorites and it may harbor clues to the origin of life. 4 Vesta is much more evolved than 1 Ceres and is one of the few asteroids we know of that may be differentiated into core, mantle, and crust much like Earth and Mars.

Dawn launched on September 27, 2007. It will reach Mars in early 2009 where it will gravity assist to send the spacecraft into the asteroid belt. Dawn will reach the belt in October 2011 and will spend ~6 months orbiting 4 Vesta, conducting science along the way. Then it will leave 4 Vesta and travel to 1 Ceres arriving there in August 2015. This is the first time that a single spacecraft has been designed to travel to two different worlds in succession, orbiting each in turn. Each will be thoroughly photographed and maps made with an infra-red mapping spectrometer to determine their precise chemical and mineral composition.

Manned Missions

Now that NASA is talking about returning to the Moon by 2020, there is also discussion about a plan to send a manned mission to an asteroid using the same Orion capsule and Ares rocket that will be used for lunar missions. The obstacles to such a venture are great, of course, but solving them and many more will be a necessary part of any future long-term space explorations by humans. For many years, there has been tremendous interest in asteroids and speculation about the feasibility of mining them for the raw materials necessary to make space colonization a reality. Until that day comes, we will have to continue to wait for the asteroids to come to us as meteorites, as they have been doing since Earth was young.

References and Useful Web Sites

Books

Beatty JK, Petersen CC, Chaikin A. *The New Solar System, Fourth edition*. Sky Publishing Company; 1999.
Cunningham CJ. *Introduction to Asteroids*. Willmann-Bell; 1988.
Bell J, Mitton J (editors). *Asteroid Rendezvous NEAR Shoemaker's Adventures* at Eros. Cambridge University Press; 2002.
Lewis JS. *Rain of Iron and Ice*. Helix Books; 1996.

Web sites

Small Bodies Data Archives, University of Maryland College Park—http://pdssbn.astro.umd.edu
Lunar and Planetary Science links to asteroid information—http://nssdc.gsfc.nasa.gov/planetary/planets/asteroidpage.html

CHAPTER THREE

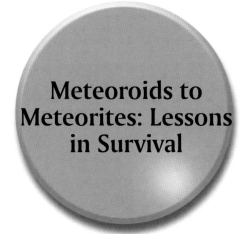

Meteoroids to Meteorites: Lessons in Survival

Atmospheric Entry

Meteorites are rocks from other worlds. To reach Earth intact, all *meteoroids* must pass a rigorous test before they can become *meteorites*. They must survive passage through Earth's dense atmosphere. Earth's atmosphere provides an effective shield against most incoming meteoroids. With this protective shield in place, they stand little chance of reaching Earth's surface without considerable damage. Particles from dust-sized to two or three millimeters diameter normally don't make it. They are totally consumed, ablated away by frictional heating in the atmosphere. Larger bodies fortunately can survive their passage relatively unscathed, though reduced in mass and size but still intact. Every year, worldwide, about 40,000 tons of meteoritic debris make it through the atmosphere. Among them are the meteorites we see preserved in museums and in private collections. They are the ones that have passed the fiery test.

One should not confuse the occasional bright meteor with the most spectacular of meteors called *fireballs* or bolides. They are produced by relatively large meteoroids ranging from walnut-sized to several feet across. By long established convention, a fireball is any meteoroid that has attained an apparent magnitude of −5 or brighter with no real upper limit (the Sun is −26.5). The brightest fireballs often have magnitudes exceeding the brightness of the full moon (−12.5). These chunks of rock are often large enough to survive atmospheric passage. They are pieces of asteroids that have been involved in impacts with other asteroids in space numerous times in their history. Often the most energetic impacts result in fractures that structurally weaken them. If they happen to encounter Earth's atmosphere in their weakened condition they stand a good chance of fragmenting into several pieces.

There are a number of physical effects that are common to most if not all fireballs. Some examples appear in the following historical accounts in O. C. Farrington's 1915 self-published book, *Meteorites*:

> At the fall of Tabory, Perm, Russia, which took place at 12:30 P.M., August 30, 1847 a fiery mass appeared in a clear sky and moved in an almost horizontal direction toward the northeast. It spread sparks in its way which left a bright smoky trace after them, and some observers saw an illuminated stripe remaining

after the mass had passed. The fiery mass remained in view only two or three seconds. Two or three minutes later, sounds like the firing of many cannon were heard. In several villages of the region, black, warm stones weighing from two to twenty pounds fell to the Earth.

The fall at Hessle, Sweden, which occurred, January 1, 1869, at 12: 20 P.M. was accompanied by a sound resembling heavy peals of thunder, followed by a rattling noise as of wagons at a gallop and ending with a sound at first like an organ tone and later like that of hissing. Many small stones fell in this shower. One struck ice close to where a man was fishing and rebounded. He picked it up and found it warm.

At the fall of the iron meteorite of Braunau, Bohemia, which took place July 14, 1847, at 3:45 A.M., the people of Braunau were wakened from sleep by two violent sounds like cannon shots followed by a whistling and rushing sound which lasted several minutes. Those who hastened into the open air saw to the northwest in a sky in which some stars were yet visible, a small, black cloud. This cloud glowed and emitted tongues of light, two of which flashed to the Earth. About the fiery cloud was seen one of ash-gray color which finally disappeared in the direction in which the wind was blowing. An iron meteorite weighing 48 pounds was found in a hole three feet deep, and this, six hours after the fall was so hot as to burn the hands of those who touched it. About a mile away to the southeast a mass weighing 35 pounds fell through the roof of a house and near a bed where three children were sleeping.

Fireballs

All these stories share similar descriptions of light, heat, and sound. The fireball stage is a real sound and light show. It is physics in action. All bodies have *inertia*. Inertia is simply a body's resistance to a state of change in its motion. If a body is at rest, it will remain at rest or, if in motion, will remain in motion traveling a straight line until compelled to change that state by the action of an exterior force. The force required to change the state of motion is the *momentum* of the body. Momentum is the product of the body's mass (the amount of material it contains) and its velocity. Thus, momentum = mass × velocity. A large meteoroid hitting the top of the atmosphere at a typical entry velocity of 25 miles per second has enormous momentum. The force needed to slow it down comes from the atmosphere itself. The atmosphere reduces the meteoroid's momentum by creating a drag on it, and, at the same time, slows its velocity. Further reduction of its momentum occurs when the meteoroid is instantaneously heated to the point of melting, causing it to lose mass by ablation. Like momentum, a moving mass also has energy associated with its motion, referred to as *kinetic energy*. Mathematically, the kinetic energy equation is similar to the momentum equation and looks like this: $KE = \frac{1}{2}mv^2$. Here, velocity is the more important factor because the kinetic energy varies with the *square* of the velocity and only the first power of the mass. If two meteoroids have the same velocity but one has twice the mass of the other, the more massive body will have twice the kinetic energy of the other. If two meteoroids have the same mass but velocities differing by a factor of two, the meteoroid with the greater velocity will have four times the kinetic energy. Here we have treated the mass as a constant. This is not necessarily true, for meteoroids passing through Earth's atmosphere can lose as much as 90% of their mass due to ablation. The mass must therefore be treated as a variable.

Light, Sound and Heat

Kinetic energy can be converted to other forms of energy such as heat and light acting together to produce a fireball. The solid meteoroid usually does not become luminous until it has fallen to an altitude of around 60 mi (100 km). At that altitude aerodynamic drag becomes an important factor, producing a substantial resistance to the meteoroid's motion. The meteoroid starts to convert some of its kinetic energy to heat which begins to melt the outer surface of the meteoroid. At the same time, the meteoroid begins to give off a weak light as the temperature climbs to over 1500° C. The light of a fireball is produced by two different mechanisms operating simultaneously. First, the solid body of the meteoroid becomes incandescent as the melting point is rapidly reached. This, by itself, gives off light but not yet a sufficient amount of light to be easily seen from the ground. But as heating continues, the air surrounding the meteoroid begins to be heated simultaneously with the meteoroid. Atmospheric atoms surrounding the forming fireball begin to ionize (lose electrons). Almost immediately the atmospheric atoms recapture their electrons, releasing light at the same time and causing the air around the meteoroid to become incandescent. This process produces a huge glowing spherical air mass measuring hundreds of feet across. This is the fireball we see from the ground.

The sound of a fireball is an altogether different experience. It is an eerie moment when the fireball begins its rapid journey across the sky. Trees and tall buildings cast long moving shadows as they seem to race with the fireball. Seconds go by and not a sound is heard. Suddenly, without warning, the fireball explodes, scattering myriads of fragments that briefly maintain their courses among the stars. All of this occurs in absolute silence. Seconds and minutes go by. The fireball vanishes. Still, silence. Then, when you least expect it, a tremendous series of explosions rock the silence. The fireball's shock wave has finally arrived, announcing its presence by a series of ground-shaking sonic booms. These sounds are caused by pressure waves generated in the atmosphere by the hypersonic flight of the fireball. The light of the fireball and the sounds generated by the shock waves from dozens of meteoritic fragments cannot occur simultaneously. Light is propagated over that short distance (60 mi) almost instantaneously, whereas sound traveling at about 1,130 ft/s (330 m/s) lags far behind. Depending upon the distance of the fireball from the observer, there could be a difference of between 30 seconds to several minutes delay after the passage of the fireball.

Over the last decade or two there have been numerous stony meteorites that have passed through rooftops and landed on city streets where they were picked up immediately after landing. Never has there been a report of a meteorite too hot to handle. Warm, yes; hot, no. The reason is simple. At an average altitude of about 50,000 ft, the meteorite's cosmic speed has been reduced to zero; it is now simply subject to the laws of gravity which maintains its fall at a few hundred miles per hour, too slow to produce compression of the atmosphere with resultant heating. Only the outer millimeter or so is actually affected by the melting process and is rapidly ablated away. No appreciable heat is conducted into the interior of the meteorite. The temperature at 50,000-ft altitude is about −50 °F. This low temperature aids in rapidly chilling the falling rock. Long before hitting the ground the meteorite's surface temperature has been reduced to between lukewarm and stone cold. The meteorite may even be coated with a thin layer of ice. In fact, some meteorites have been found minutes after landing, resting on top of a snow bank—without melting the snow.

Iron meteorites may be something else entirely. Iron is a much better heat conductor, and occasionally we see a heated zone up to a few millimeters thick around the perimeter of a cut iron slab, evidence of severe heating. Still, the short duration of flight constrains the heating to this depth at most. When iron meteorites reach Earth they could be uncomfortably hot to the touch but not glowing red hot.

Ablation

By far, the most destructive process occurs during the incandescent phase of atmospheric entry. But the ablation process is a double-edged sword. The cooling effect of ablation is a necessary mechanism that enables the meteoroid to survive its passage to Earth's surface. In other words, it must ablate to survive. As the drag on the meteoroid increases rapidly with descent in the atmosphere, the forward end heats to incandescence, melts, and begins to rapidly lose mass. The expelled mass carries with it the heat of vaporization. This cools the body by carrying away the heated portion and inhibiting the flow of heat to the interior. The melted material flows into the air stream behind the fireball forming a long narrow dust train. This dust is composed of tiny microscopic liquid droplets that quickly solidify.

Meteoroids with high initial entry velocities are subject to far greater atmospheric resistance than slower-moving meteoroids and, as a result, suffer greater ablative effects. This is shown in Figure 3.1 in which entry velocities of a one ton iron meteoroid are related to the percent of the initial mass retained upon reaching Earth's surface. Initial velocities of 12 and 24 mi/s (19 and 39 km/s) at an entry angles of 45° are compared. Note that the higher velocity body suffers much greater mass loss, retaining only about 55% of its initial mass. By comparison, the slower moving body retains 86%.

The fall of the Sikhote-Alin meteorite is a classic example of a witnessed fall of a large iron meteorite. On the morning of February 12, 1947, an enormous fireball appeared over the Sikhote-Alin mountains in eastern Siberia leaving a train of vaporized and recondensed material estimated to weigh over 200 tons, the most massive ever witnessed. The Russian artist, P.I. Medvedev, from the village of Iman in the Maritime Province of Eastern Siberia had just set up his easel pointing eastward over the Sikhote-Alin Mountains. At 10:38 a.m. the fireball appeared, traveling from north to south heading toward the mountains. Medvedev, an eye-witness to the spectacular event, managed to sketch the fireball and long train immediately (Figure 3.2). Thousands of beautifully sculpted iron meteorites pelted the ground, making over 100 impact craters and penetration holes, the largest being 87 ft across and nearly 20-ft deep from rim to floor. The remaining material that reached the ground was estimated to be about 70 tons. After more than 50 years of searching, over 25 tons of meteoritic material have been found at the impact site. Today, Sikhote-Alin meteorites are highly prized in public and private collections throughout the world (Figure 3.3).

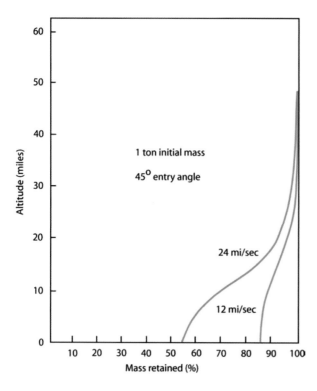

Figure 3.1. The entry velocity of a 1-ton iron mass determines the amount of the initial mass that will reach Earth's surface.

Figure 3.2. Ten years after the Sikhote-Alin fall, the Soviet Union issued this postage stamp commemorating the event, as depicted by the artist who witnessed it.

Figure 3.3. A typical Sikhote-Alin meteorite with characteristic regmaglypts produced by ablation. This one measures 11 cm.

Brecciated Meteorites and Multiple Falls

Nearly all meteorites show moderate to severe signs of shock and *brecciation* (breakage of rock into smaller fragments). The stony meteorites in particular often break into many fragments while still in space. This is usually the result of impacts between their associated asteroid parent bodies and other asteroids early in the history of the Solar System. In some cases material that has been liquefied or partially liquefied by shock pressures in excess of 90 GPa (greater than 13,000,000 lb/in.[2]) forms an *impact melt breccia*. This pressure is substantial enough to produce wide-spread melting of the rock within and around the floors of newly produced impact craters. Instantly, at impact, a melt lens forms a kilometer or more beneath the crater floor. The rock loses its original texture as it is partially melted in the lens. The melting is usually incomplete however, and the result is a mix of melted rock and unmelted pieces (clasts). Impact melt breccias are most often seen in chondritic meteorites since there are more of them. A good example is the impact melt breccia seen in Figure 3.4. Here, islands of chondritic material are surrounded by almost featureless matrix.

Sometimes during impact the entire meteoroid fractures but amazingly does not shatter. Instead, the fragments stay together while in space. In this weakened condition, however, they often break into pieces as they enter Earth's atmosphere. A classic example of the breakup of a stony meteoroid occurred on October 9, 1992, over Peekskill, New York. It was football season in the high schools along the eastern seaboard and the games were being played. Just minutes before 8:00 p.m. a brilliant fireball sliced across the sky. It took only seconds to traverse eastern Kentucky, along its north-north easterly flight path, then pass in succession over North Carolina, Maryland, and New Jersey producing delayed sonic booms along the way. At the midway point in its path, the meteoroid began to fragment, breaking into more than a dozen pieces, each fragment forming its own fireball and trail (Figure 3.5). The main mass of the Peekskill stone weighed 12.6 kg and was located immediately after it fell. It had fallen through the back end of a parked car barely missing the gas tank. Sixteen-year-old Michelle Knapp (owner of the car) heard a crash outside her house and, upon investigating, discovered the meteorite in a shallow impact pit beneath the car's trunk. When the meteorite was cut to reveal its interior it was no surprise to find it was a brecciated chondrite. Shock veins crisscrossed the fractured mass. The meteorite itself was composed

Figure 3.4. An example of an impact melt breccia. This specimen was supplied by Bruno Fectay and Carine Bidaut, The Earth's Memory, meteorite.fr.

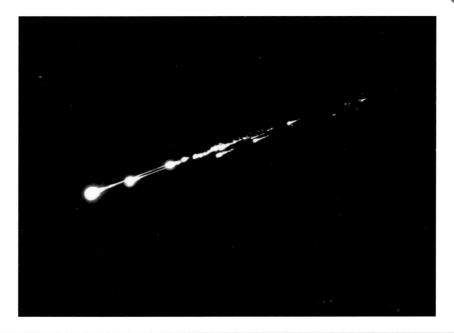

Figure 3.5. The Peekskill, New York, multiple fireball shortly after fragmentation on October 9, 1992. Courtesy the *Altoona Mirror*.

of many angular fragments that fit together like a puzzle. Dark veins of glass and recrystallized minerals permeated the fragments holding them together. This common texture is called a *monomict breccia* (Figure 3.6a–c).

Some meteoritic breccias are composed of two or more fragments differing in texture and composition from the host rock. These are called *polymict breccias*. A good example is the achondrite (aubrite) Cumberland Falls (Figure 3.7). The light clasts are the magnesium-rich mineral, enstatite, which also composed the original pre-shocked parent body. The dark fragments have an anomalous chondritic composition intermediate between E (enstatite chondrites) and the H (ordinary chondrites). (Meteorite classification will be discussed at length in Chap. 4.) Here, material from both impacting bodies are mixed, intermingled into one rock. Occasionally, breccias are found with clasts of similar composition but of different petrographic types (ordinary chondrites types 3, 4, 5, and 6). These are called *genomict breccias*.

One final breccia type should be included, the *regolith breccia*. This breccia is formed from a layer of rock debris produced by impacts on the surface of another world such as the Moon. Nearly the entire surface of the Moon is covered by a veneer of loose soil composed of broken rock ranging from a few micrometers to tens of meters in thickness. Here, long after formation, the Moon's shallow surface layer was subjected to repeated impacts by tiny particles which created an unconsolidated layer of fragments and dust resting on solid bedrock, much as the Apollo astronauts found it more than 30 years ago. This layer is the lunar regolith breccia. Such unconsolidated layers are found on the surfaces of asteroids as well, and some samples eventually find their way to Earth.

Figure 3.8 shows the H4 chondrite from Dimmitt, Texas. The cut and polished face of the slab reveals the light/dark structure typical of an asteroidal regolith breccia with light clasts and dark matrix material. The light/dark structure is easy to distinguish when compared to other chondritic

a

b

c

Figure 3.6. a The interior of the Peekskill H6 chondrite, which is fractured throughout. It is a monomict breccia. **b** A detail showing the crust which is 1.8 mm thick. **c** The car and the meteorite. Note red paint from the car on the meteorite. Photographs by Iris Langheinrich, R. A. Langheinrich Meteorites, www.nyrockman.com.

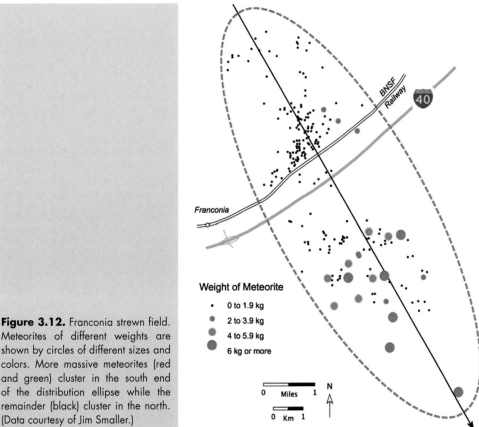

Weight of Meteorite

- 0 to 1.9 kg
- 2 to 3.9 kg
- 4 to 5.9 kg
- 6 kg or more

Figure 3.12. Franconia strewn field. Meteorites of different weights are shown by circles of different sizes and colors. More massive meteorites (red and green) cluster in the south end of the distribution ellipse while the remainder (black) cluster in the north. (Data courtesy of Jim Smaller.)

Figure 3.13. The fresh crust of this Bassikounou stone which fell on October 16, 2006, in Mauritania displays regmaglypts and breaks revealing its lighter interior. It weighs 3300 g. Cube is 1 cm on a side. Courtesy of Peter Marmet, www.marmet-meteorites.com.

A final characteristic of the fusion crust seen on many stony meteorites is the presence of contraction cracks. These cracks may look similar to the cracks we saw when we looked at meteoritic breccias. However, contraction cracks are usually much finer, being produced by the rapid cooling of the now spent stony meteorite. The cracks are generally not very deep, seldom as deep

Figure 3.14. A stone from the Norton County, Kansas aubrite fall still embedded in the dirt, now in the collection of the University of New Mexico. The light brown crust makes it difficult to find. Courtesy of Al Mitterling.

Figure 3.15. The shiny black fusion crust of an achondrite is most notable in Camel Donga eucrites like this one from Western Australia. Specimen weighs 462 g and measures 111 × 75×50 mm. Courtesy of Jim Strope, www.catchafallingstar.com.

as the thickness of the crust itself (Figure 3.16). Contraction cracks are important entrance points into the meteorite's interior where chemical and mechanical weathering begins soon after the meteorite lands. Later we will look more closely at how meteorites weather in the terrestrial environment.

Iron meteorites also form fusion crusts but they are substantially thinner than the average stony crust, being only a fraction of a millimeter thick. A fresh crust appears blue-black in color and looks like freshly-welded steel. Silicate minerals are not involved in crust-forming in irons. Iron crusts are composed almost entirely of iron oxide. Of all the fusion crusts, iron crusts are by far the most fragile. They are much more subject to chemical weathering (rusting) which can easily destroy the thin veneer.

Another effect occasionally seen on the edges of iron meteorites is heat alteration. For example, the small Sikhote-Alin in Figure 3.17 shows the result of reheating in the outer few millimeters all the way around, resulting in a granular texture which has replaced the original octahedral structure. Etching has revealed the distorted edge and the parallel-running Neumann lines in its interior. Figure 3.18 shows another dramatic example of reheating in a Henbury iron from Northern Territory, Australia.

Figure 3.16. Contraction cracks in a very fresh stony meteorite from Bensour, which fell in Morocco on February 11, 2002. Field of view is 15 mm.

Figure 3.17. This 19.5 g Sikhote-Alin iron shows the effects of heat alteration on its surface. Its octahedral structure has been completely recrystallized. Note Neumann lines. Courtesy of Martin Altmann, www.chladnis-heirs.com.

Figure 3.18. Extraordinary Henbury iron with a recrystallized rim ranging from 1 to 11 mm in width. Widmanstätten figures are preserved in the interior of the specimen. Courtesy of Mirko Graul.

The Secondary Fusion Crust

In a field of meteorites it is not unusual to find several that show signs of a *secondary fusion crust* (Figure 3.19). Usually, the ablative process builds a nearly complete smooth coating over the entire surface of a meteoroid and that surface is replaced many times while in the atmosphere. The presence of a secondary fusion crust tells us that these meteorites have broken in flight, losing some of the primary crust in the process. The freshly broken face will immediately begin to rebuild a new, thinner crust—the secondary fusion crust. This crust is always much thinner than the primary crust since it forms later in flight and often does not completely coat the newly-formed fragments. The secondary fusion crust is not as smooth as the primary crust since the broken face was subject to less ablation.

Angularity of Stones

Meteorites have definite shapes you should look for. They are the product of the ablative process and fragmentation. One would expect the final shape to be irregular since they usually explode high in the atmosphere sending fragments in all directions. The original shape while in space can never be known. However, many meteorites acquire an angularity that is definitely rectangular with near-90° angles predominating. If the meteorite develops its shape while still high in the atmosphere, the straight edges tend to be rounded rather than sharply defined. Specimens from the multiple fall of the Gao-Guenie H5 meteorites in Burkina Faso (Upper Volta) are good examples. Two meteorite showers occurred exactly one month apart in 1960, producing chemically and texturally identical meteorites (Figure. 3.20).

Regmaglypts, Flow Features and Oriented Meteorites

Another product of ablation that occurs during a meteoroid's brief atmospheric passage is the development of surface pits or *regmaglypts* on stones and irons alike. The common name for these

Figure 3.19. Secondary fusion crust on the chondritic meteorite Campos Sales, an L6 stone which fell on January 31, 1991, at 10:00 pm. in Ceara, Brazil. A shower of stones fell. This one weighs 1,494g. Photograph by Geoffrey Notkin/Aerolite.org, © The Michael Farmer Collection/www.meteoritehunter.com

is "thumb prints," since they are about the size of the human thumb. Regmaglypts in stones tend to be shallower than in irons, and therefore not as well defined. Figure 3.21 shows a regmaglypt texture in a Mbale L6 chondrite. If you compare it to the Sikhote-Alin iron texture in Figure 3.22 you see that the regmaglypts are much deeper and better defined in the iron. If an iron meteoroid breaks up explosively, the meteorites may be very distorted and may resemble bomb fragments. For this reason they are referred to as shrapnel. Some specimens may be a combination of both (Figure 3.23). The 1 kg Henbury iron has regmaglypts that have weathered to fine edged points, which are typical of this meteorite (Figure 3.24).

When the meteoroid is incandescent during its brief passage through the atmosphere it is melting and rapidly losing mass. Sometimes this results in meteorites, irons as well as stones, that display distinctive flow features. Often fine lines are seen radiating from the leading edge of the meteorite, reflecting its flight orientation in radial flow lines. Sometimes a lip forms over the trailing edge and occasionally creates a spectacular meteorite such as the one shown in Figure 3.25. This meteorite has flow features resembling teeth in a comb. This is truly an example of ablation in progress, a moment frozen in time by rapid cooling of the stone.

Figure 3.20. This Gao meteorite formed when a fragmentary piece broke along 90 degree fracture planes. It measures 7 × 5 × 4 cm.

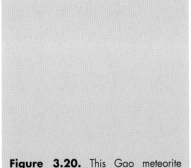

Figure 3.21. Regmaglypts or "thumb prints" on a stone. This Mbale L6 chondrite fell in Uganda in 1992.

Figure 3.22. Regmaglypts on a Sikhote-Alin iron. This beautiful specimen weighs 9.4 kg. Courtesy of Jim Strope, www.catchafallingstar.com.

Figure 3.23. A 20 lb Gibeon iron from Namibia with typical regmaglypts and very unusual shock deformed structures. Courtesy of Howard Wells.

Figure 3.24. This Henbury 1 kg iron from Northern Territory, Australia has very sharply defined regmaglypts. The color is typical of Henbury irons and comes from the red laterite soil in which the meteorite was embedded. Courtesy of Dr. Svend Buhl www.meteorite-recon.com.

Figure 3.25. Flow structures on Dhofar 182, a eucrite found in Oman in 2000. The stone was oriented in flight which formed the radial flow lines. Rapid cooling preserved the flowing crust. Courtesy of Jörn Koblitz, www.metbase.de.

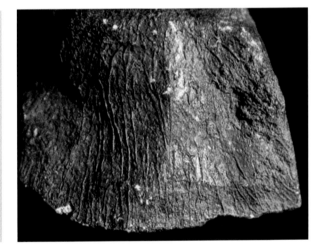

Figure 3.26. Delicate flow features on the crust of a Millbillillie eucrite achondrite seen here under special lighting. This specimen is 2.5 cm.

To reveal these delicate features, the meteorite crust must be carefully cleaned and free of chemical weathering. In the case of the Millbillillie meteorite in Figure 3.26, a special fiber optic focused light with an intense narrow beam was placed along the edge of the specimen highlighting the delicate "rivulets" and emphasizing equally delicate shadows. These flow structures can be very subtle, but an exceptional example is seen in the Lafayette stone which displays distinct radial flow lines (Figure 3.27). Irons can also demonstrate incredibly fine flow structures like those seen in the Sikhote Alin meteorite in Figure 3.28.

By far, the most dramatic structures can form when meteoroids become oriented during atmospheric flight. The vast majority tumble uncontrollably which tends to smooth them out, creating roughly spherical shapes. But if a meteoroid acquires maximum drag force, it stabilizes and is prevented from tumbling. In that case it continues its rotation along the direction of motion

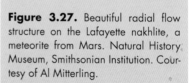

Figure 3.27. Beautiful radial flow structure on the Lafayette nakhlite, a meteorite from Mars. Natural History Museum, Smithsonian Institution. Courtesy of Al Mitterling.

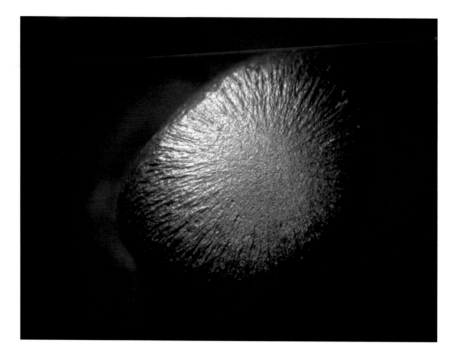

Figure 3.28. Fine flow structures on a tiny Sikhote-Alin (field of view is 14 mm). The club-shaped meteorite is only 21 mm long and weighs 8 g. Specimen from Tom Smith of Magic Mountain Gems and Meteorites.

which results in a cone or shield-shaped specimen. Figure 3.29 shows the evolution of an oriented meteorite.

In 1886, about 6 mi (10 km) east of Cabin Creek, Arkansas an extraordinary fall of a 107 lb (48.5 kg) oriented iron meteorite was witnessed and recovered. O. C. Farrington tells us:

> *The Cabin Creek, Arkansas, meteorite, one of the few irons and the largest iron ever seen to fall, fell at 3:00 P.M., March 27, 1886. It gave the first indication of its approach to the party who was nearest it, a lady in*

Table 3.1. A list of the largest iron meteorites by weight in metric tons

	Largest iron meteorites			
Meteorite	Year and place of discovery	Weight (metric tons)	Structural class	Chemical class
1. Hoba	Namibia, 1920	60	Ataxite	IVB
2. Campo del Cielo	Chaco, Argentina, 1969	37	Coarse octahedrite	IAB-MG
3. Cape York (Ahnighito)	West Greenland, Greenland 1894	31	Medium octahedrite	IIIAB
4. Armanty	Xinjiang, China	28	Medium octahedrite	IIIE
5. Bacubirito	Sinaloa, Mexico, 1863	22	Finest octahedrite	IRUNGR
6. Cape York (Agpalilik)	West Greenland Greenland, 1963	20	Medium octahedrite	IIIAB
7. Mbosi	Rungwe, Tanzania, 1930	16	Medium octahedrite	IRUNGR
8. Campo del Cielo	Chaco, Argentina, 2005	15	Coarse octahedrite	IAB-MG
9. Willamette	Oregon, USA, 1902	14	Medium octahedrite	IIIAB
10. Chupaderos I	Chihuahua, Mexico 1852	14	Medium octahedrite	IIIAB
11. Mundrabilla I	Western Australia, Australia 1966	11.5	Medium octahedrite	IAB-UNGR
12. Morito	Chihuahua, Mexico, 1600	11	Medium octahedrite	IIIAB

Figure 3.34. Tim Heitz atop the 37-ton Campo iron. It is the second largest iron and rests today near its original location in the Campo del Cielo strewn field. Courtesy of Tim Heitz, Midwest Meteorites.

when the Spanish governor was told about them by local natives. They said the pieces of iron had fallen from heaven and the area where they were found was known as Campo del Cielo—field of the sky, or field of heaven. The first large mass discovered was called Meson de Fierro (large table of iron). It was too big to be transported and subsequently its location has been lost. But many masses have been found, large and small, and the Campo irons today are some of the most plentiful and attractive on the market. Though they have been on the ground for about 4000 years, and some are prone to rust, others are quite stable. The original iron asteroidal body contained silicate inclusions, which can be seen in some of the specimens. Whatever the size, Campo irons should be a welcome addition to any meteorite collection.

Weathering of Meteorites

Mechanical Weathering

For billions of years meteorites in the environment of space have resisted most weathering. They are far removed from the damaging effects of water and oxygen. But Earth is an alien environment for a meteorite. Under normal terrestrial conditions they cannot survive long after impact without measures being taken to preserve them. There are two basic types of weathering: *mechanical* and *chemical*, and meteorites are subject to both. Mechanical weathering actually commences when meteorites suffer one or more episodes of fragmentation while still in space. They further fragment when they impact Earth's surface while traveling hundreds of miles per hour. Mechanical weathering continues breaking a meteorite apart through wind abrasion, transportation by water, temperature extremes, and even the activity of plants and animals. The glassy fusion crust provides some protection on those sides that have not fragmented (unusual), but contraction cracks often begin to form, allowing water into the meteorite's interior where the far more destructive chemical weathering can begin. If a meteorite lands in an area that is subject to a freeze/thaw cycle, this could be sufficient to widen the contraction cracks through ice wedging.

Chemical Weathering

Most meteorites are finds, not falls and as such have been exposed to the elements, perhaps for millennia. The main chemical weathering reactions are oxidation, hydration and solution. In time, new minerals will be produced. On a stone, the dark crust can lighten to a medium brown as the iron in the meteorite oxidizes to a new weathering mineral like goethite. Olivine and feldspar turn into clay-like minerals. Iron meteorites are especially prone to rusting along kamacite plate boundaries (Figures 3.35 a and b and 3.36).

Heavily weathered stony meteorites often have thick rinds replacing the original fusion crust. For example, two meteorites from Gold Basin, Arizona illustrate the effects of weathering on the exterior surface (Figure 3.37a and b). The stone on the left was found about six inches below ground level with its fusion crust completely replaced by a thick rind. The stone on the right has a thin, black fusion crust on the portion of the exterior surface that was positioned above ground level. Chemical weathering can strip away a fusion crust in less than a few centuries. The fusion crust on the Gold Basin L4 chondrite will all but disappear in only about 12,000 years.

The interiors of most ordinary chondrites commonly contain a uniform distribution of iron-nickel metal grains. These grains rapidly oxidize in the interior resulting in the formation of patches of limonite, an amorphous, hydrated iron oxide (rust) that effectively stains the primary olivine and pyroxene minerals (Figure. 3.38).

Figure 3.35. A Campo del Cielo iron showing rust along the plate boundaries, before (**a**) and after (**b**) cleaning. This specimen measures 6.5 cm.

a

b

Figure 3.36. Photomicrograph of a Toluca iron with blebs of iron oxides and hydrochloric acid seeping from between the plates. In time this process can destroy a meteorite. Field of view is 9 mm.

a

b

Figure 3.37. Weathering on the crust of two Gold Basin L4 chondrites. **a** The crust on this completely buried 240 g stone has been totally replaced by a thick rind. **b** Crust remains on a 53-g stone which was partly above ground.

Many methods have been devised to prevent chemical weathering in meteorites, especially in irons. For example, after etching an iron meteorite slab to bring out the Widmanstätten figures using dilute nitric acid as the etchant, the specimen is neutralized, washed with water, oven dried and soaked in 99% alcohol to dry it out. Then it is protected from the outside air by coating it with an acrylic finish. If the meteorite contains chlorine, however, it will most likely begin the cycle anew.

In 1993, the meteoriticist F. Wlotzka of the Max Planck Institute in Mainz, Germany developed a weathering scale for ordinary chondrites. He detailed six weathering grades on polished sections and in thin sections and designated seven progressive weathering states labeled W0 to W6. We will look at the effects of weathering on meteorites under the petrographic microscope in Chap. 11.

References and Useful Web Sites

Books

Bevan A, de Laeter J. *Meteorites A Journey Through Space and Time*. Smithsonian Institution Press; 2002.
Grady M. *Catalogue of Meteorites*. 5th edn. Cambridge University Press; 2000.
McSween H Jr. *Meteorites and Their Parent Plants*. 2nd edn. Cambridge University Press; 1999.
Norton OR. *Rocks From Space*. 2nd edn. Mountain Press; 1998.
Norton OR. *Cambridge Encyclopedia of Meteorites*. Cambridge University Press; 2002.
Reynolds MD. *Falling Stars: A Guide to Meteors and Meteorites*. Stackpole Books; 2001.

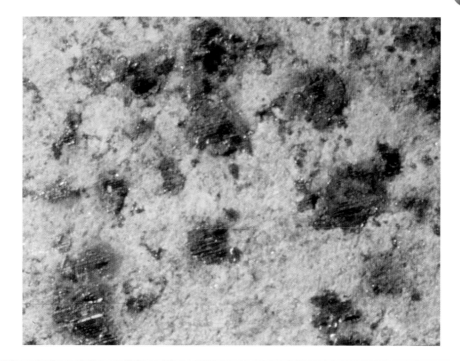

Figure 3.38. Photomicrograph of Bruderheim L6 chondrite. The cut surface shows limonite staining of the matrix as it forms around the iron-nickel grains. Field of view is 9 mm.

Sears D. *The Origin of Chondrules and Chondrites*. Cambridge University Press; 2004.
Zanda B, Rotaru M. *Meteorites Their Impact on Science and History*. Cambridge University Press; 2002.

Magazines

Meteorite Magazine: www.meteoritemag.uark.edu/index.htm
Meteorite Times online magazine: www.meteorite-times.com

Useful web sites with information

About meteorites and collecting
International Meteorite Collectors Association: www.imca.cc
www.meteorites4sale.net
www.meteorite.com
www.meteoritecentral.com
www.meteorites.com.au
www.spacerocksinc.com

For preserving iron meteorites
www.paleobond.com
www.meteorites.com.au/odds&ends/ironrust.html
http://earthsci.org/fossils/space/craters/met/met.html
www.alaska.net/~meteor/hobby.com

Part II

The Family of Meteorites

CHAPTER FOUR

The Chondrites

There was a time when scientists divided meteorites into stones, irons, and stony irons. In a general way this is still useful, though archaic. It is easy enough to tell the difference between the three types. Irons are heavier than any terrestrial rock of the same size and they are strongly attracted to a magnet. At first glance, meteoritic stones appear very much like rounded river rocks and they are composed of minerals we find in igneous rocks on Earth. Stony-irons, as the name implies, are mixtures of both. But confusion soon enters. Placing a magnet against a stony meteorite we find it too has an attraction, though not as much as the irons. Now we must be more observant. Cutting the stone to expose its interior reveals tiny silver-colored flakes of metal (Figure 4.1). We call this elemental iron, meaning iron in its most reduced chemical state. This is definitely a surprise.

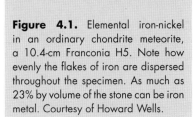

Figure 4.1. Elemental iron-nickel in an ordinary chondrite meteorite, a 10.4-cm Franconia H5. Note how evenly the flakes of iron are dispersed throughout the specimen. As much as 23% by volume of the stone can be iron metal. Courtesy of Howard Wells.

Elemental iron is almost never found in terrestrial crustal rocks. Why? Because iron rusts; that is, it oxidizes rapidly in the presence of oxygen and water, and Earth has plenty of both. Along with the surprise metal, a closer look reveals a curious texture found in no terrestrial rocks. (Texture, geologically speaking, is the general appearance of a rock indicated by its grain size, shape, and internal pattern.) Scattered throughout the matrix of the meteorite are submillimeter to millimeter-sized spherical inclusions made of very hard yellowish crystallized minerals. These spheres are called chondrules and the rock that contains them, chondrites. The minerals that make up these spherical bodies as well as the matrix that holds them in place are quite common, being found in Earth's crust and mantle. But taken as a whole, chondritic meteorites are alien to Earth. It is the mineral composition as well as the texture that determines the rock's origin, either terrestrial or "celestial." We now have the idea that these stones, unlike their cousins the irons and stony-irons, have a mineral and textural diversity found only in chondrites that will help us place them into an orderly classification scheme. We will look at that classification shortly but first we need to arm ourselves with some mineralogy.

Primary Minerals in Chondritic Stony Meteorites

As of 2007, there were more than 4,000 named minerals accepted by the IMA Commission on New Minerals and Mineral Names. Of these, about 280 minerals are known to exist in meteorites. This is only about 7% of the minerals that form in the terrestrial environment. Crustal rocks on Earth as well as on the other terrestrial planets are made of aggregates of minerals called rock forming minerals. Amazingly, these minerals are various combinations of only eight different chemical elements. They are in order of abundance: oxygen (O), silicon (Si), sodium (Na), calcium (Ca), potassium (K), aluminum (Al), magnesium (Mg), and iron (Fe). The 54 most common minerals found in meteorites are listed in Appendix 1. In the crustal rocks of Earth and in meteorites, the two most abundant elements combine to form an atomic structure called the silicon-oxygen tetrahedron, the building block of all silicate minerals (Figure 4.2). The silicon-oxygen tetrahedron (also called the silica tetrahedron) has four oxygen atoms surrounding a small

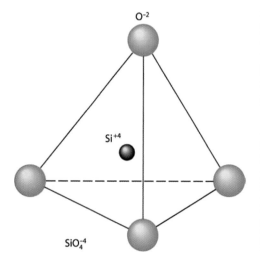

Figure 4.2. The silicon-oxygen tetrahedron. All silicate minerals are constructed with these tetrahedras. The oxygens of the tetrahedra can be chemically combined with each other to form a string of tetrahedra, or they can combine with metals like iron and magnesium. If no metal is present, the resulting mineral will be quartz.

silicon atom nestled at the center of a triangular pyramid. By itself the tetrahedron is an ion, and its chemical formula is SiO_4^{-4} with the superscript (-4) representing a negative charge on each of the oxygen atoms. When metals are present, such as iron and magnesium, the metal atoms are electrically attracted to the oxygen atoms and act as bridges between tetrahedra. Different amounts and proportions of metal atoms among the silica tetrahedra produce a variety of three-dimensional crystal lattices including those of the olivine and pyroxene minerals.

Olivine

Of all the minerals found in meteorites, three stand out as most abundant and most important. The first is olivine. Olivine is an olive green to yellowish silicate mineral found abundantly in terrestrial mafic rocks such as basalt. It is composed of varying amounts of iron and magnesium in combination with the silicon-oxygen tetrahedra. Its chemical formula looks like this: $(Fe,Mg)_2SiO_4$. Olivine is actually a group of minerals with similar structures and compositions. The $(Mg)^{2+}$ and $(Fe)^{2+}$ ions have about the same atomic size, allowing them to substitute for each other in the crystal lattice. The superscript ($2+$) means that the atoms lack two electrons and are therefore positively charged. They can readily combine with the silicon-oxygen tetrahedron since it is negatively charged. The relative amounts of magnesium and iron determine the type of olivine crystallizing out of a magma. This range of ionic substitutions is called a solid solution. There is a continuous solid solution series from magnesium-rich forsterite, or Fs, (Mg_2SiO_4) to iron-rich fayalite, or Fa, (Fe_2SiO_4). We will see later in this chapter that the relative amount of iron, both elemental and combined, determines the chemical classification of the ordinary chondrites and separates them into groups which have a limited range of compositions. They likely formed on the same asteroid parent body.

Pyroxenes

These silicate minerals are similar to olivines in that they both are solid solutions. However, pyroxene contains variable amounts of calcium in addition to iron and magnesium. The ordinary chondrites contain primarily the low calcium orthopyroxene, enstatite $(MgSiO_3)$. The greatest difference between the olivines and pyroxenes is that the olivines have more metal than pyroxenes. With less metal in the pyroxenes, the silicon-oxygen tetrahedra are forced to share some of their oxygens. (If no metals are present the mineral will be quartz (SiO_2) where the SiO_4^{-4} ion shares all oxygen atoms of adjacent tetrahedra.)

Iron-Nickel Minerals

Perhaps the most surprising primary mineral found in ordinary chondrites is elemental iron-nickel (FeNi), hereafter referred to as metal. We saw earlier that when an ordinary chondrite is cut and polished to reveal its interior composition and texture, the metal is immediately obvious as bright, star-like metallic grains against the backdrop of the matrix. A good magnet will immediately reveal its iron composition. Actually, the metal is more than just iron. It is an iron-nickel alloy. The iron is always alloyed with nickel in meteorites. The nickel content can be as little as 5% or as much as 25%. In the field, the presence of FeNi metal in a "terrestrial" rock is a nearly certain sign of a stony meteorite. As much as 23% of the total iron can be in the elemental state. The remaining iron is found in combined form as iron oxides, sulfides, carbides, phosphides, or combined in olivines and pyroxenes.

Accessory Minerals

Troilite

Common stony meteorites contain accessory or small amounts of a sulfide of iron, written FeS. In meteorite circles it is usually referred to as troilite. It was named after the eighteenth-century Italian Jesuit priest and meteorite investigator, Father Domenico Troili. Troilite is usually easy to distinguish with the eye alone as it normally has a bronze-like color that stands out, especially compared to the silvery colored iron-nickel alloy that is always present in some amount in all stony meteorites. Often the two are seen joined together. Troilite appears very similar to the iron sulfide pyrrhotite, commonly found in terrestrial rocks. The primary difference is that pyrrhotite has magnetic properties whereas troilite does not. Many iron meteorites have nodules of troilite, often surrounded by graphite.

Iron Oxide

The most common oxide of iron in meteorites is magnetite. It is a major component of the fusion crust that forms around stony meteorites as they burn up entering Earth's atmosphere. Magnetite is also found in the matrices of carbonaceous chondrites. Magnetite is strongly magnetic.

Plagioclase Feldspar

Feldspars are common minerals found in the Earth's crust. They are found in accessory amounts in most stony meteorites and mesosiderites (stony-irons) but in large amounts in basaltic achondrites. Plagioclase feldspars form by solid solution with variable proportions of sodium and calcium ions.

Elemental Abundances in Chondritic Meteorites

The chondrites are considered the most primitive of all the meteorites. Their elemental compositions are very close to that of the Sun. By mass, the Sun is about 73.5% hydrogen and 25% helium. If we remove these gases and the remaining volatile components such as oxygen, nitrogen, carbon, and neon, we are left with only about 1.5% for the remaining nonvolatile elements. The elemental composition of the most primitive meteorites, the CI carbonaceous chondrites, is usually compared to abundances in the solar photosphere. The graph in Figure 4.3 compares elemental abundance of the Sun plotted against the elemental abundance of CI carbonaceous chondrites. The solid line is the solar abundance of nonvolatile elements. Note that the CI elemental abundances of nonvolatile elements lie very close to this line, demonstrating that the primitive chondrites closely match the Sun's elemental composition.

Chemical Types of Ordinary Chondrites

Of all the stony meteorites observed to fall to Earth, the vast majority (85%) are ordinary chondrites. This curious name should not imply that chondrites are "ordinary" in the usual sense of the word. The name chondrite is derived from the presence of chondrules, tiny bodies of igneous

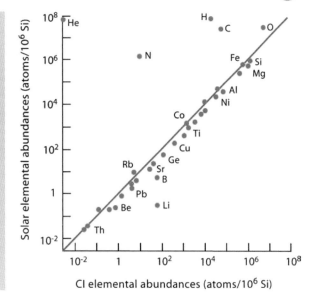

Figure 4.3. Graph comparing elemental abundance of the Sun plotted against elemental abundance of CI chondrites, the most primitive chondrites. The solid diagonal line is the solar abundance of non-volatile elements.

origin found in the interiors of these meteorites. All chondrites, with the single exception of the CI carbonaceous chondrites contain chondrules. They are roughly spherical in shape and have diameters between 0.1 and 4 mm with a few reaching centimeter size (Figure 4.4).

Chondrites are divided into classes, groups and clans. The classes include ordinary chondrites (OC), carbonaceous chondrites (C) and enstatite chondrites (E) (Figure 4.5). In this chapter we will explore the ordinary chondrites which are divided into three chemical groups, all having a limited range of compositions, possibly all forming on the same asteroid parent body.

By far, the ordinary chondrites are the largest class of stony meteorites, accounting for more than half of all meteorites known, both falls and finds together. Their composition is primarily silicate minerals we mentioned earlier: olivine and pyroxene plus iron-nickel metal. The metal, both elemental and combined (oxidized), is used to classify the ordinary chondrites into three distinct chemical groups: H, L, and LL chondrites. The "H" stands for "high iron." This group contains 25–30% total iron by weight. Of this, between 15 and 19 wt.% of the iron is in the uncombined elemental state with the remainder chemically bound to the silicates. Of the three, the H-chondrites, which have the greatest amount of metal, are most strongly attracted to a magnet (Figure 4.6a). The iron in a cut slab will show a bright sheen even without polishing. The metal is uniformly distributed throughout the meteorite and as such is easily distinguished from the other two groups. Besides the elemental and oxidized iron, the H chondrites are further distinguished chemically by the composition of their olivines. For H chondrites the range of composition is Fa_{15-19}. What does this mean? This tells us the proportions of magnesium and iron found in the olivine crystals of H chondrites. Pure iron-rich olivine is called fayalite (Fe_2SiO_4) and pure magnesium-rich olivine is called forsterite (Mg_2SiO_4). Most olivine crystals contain both. The olivines of H chondrites contain 15–19 mole% Fa (fayalite), or put in a less commonly used way, they contain 81–85 mole% Fo (forsterite). Clearly, the olivines of H chondrites (and indeed all chondrites) are magnesium-rich. Today, researchers determine these proportions on sophisticated equipment usually found at universities.

The second group of ordinary chondrites is the L chondrites, the "L" standing for "low iron." These intermediate ordinary chondrites have between 20 and 25 wt.% total iron, almost as much as the H chondrites. But the amount of metal is much lower, between 1 and 10 wt.%. This is obvious when one looks at the polished face of an L chondrite. The iron flakes are much reduced in number compared to the H chondrites and likewise the meteorite is not so strongly attracted to a

Figure 4.4. A field of chondrules in a slab of the Marlow L5 ordinary chondrite seen in plane-polarized light. The largest chondrule is about 1 mm in diameter.

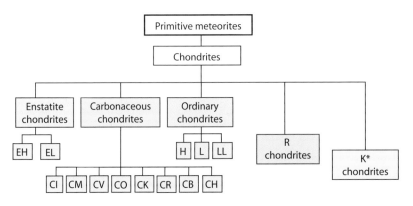

* The K chondrites constitute a grouplet of only two members

Figure 4.5. The chondritic meteorites are subdivided into three major classes: Enstatite, Carbonaceous, and Ordinary. Rumuruti (R) chondrites make up a group and K chondrites form a "grouplet" since there are fewer than 5 distinct members.

magnet. (Magnetic attraction can be very important when testing a rock in the field. Only a few terrestrial rocks are naturally attracted to a magnet while virtually every ordinary chondrite has some magnetic attraction due to the presence of elemental iron.) The composition of olivine is Fa_{21-25}; that is, the fayalite content is between 21 and 25 mole%, showing that more iron has been oxidized compared to the H chondrites. Of the three ordinary chondrite groups, the L chondrites are most commonly seen to fall, amounting to 46% of the total.

The final group of ordinary chondrites, the "LL" chondrites, are the least likely of the three groups to be seen to fall. The "LL" stands for "low metal, low total iron." Once again this reduced metal is relatively easy to see without optical aid and can still be attracted to a powerful magnet (Figure 4.6b). But compared to the "L" chondrites, the metal in "LL" chondrites is sparse indeed. The iron as metal reaches the lowest amount, between < 1 to about 3 wt.%. The total combined iron is between 19 and 22 wt.%. The fayalite content is the highest, Fa_{26-32}, between 26 and 32 mole%. These useful numbers are summarized in Table 4.1.

We have seen now that the ordinary chondrites can be classified according to the amount of total iron they contain, both elemental (metal) and chemically combined. Looking carefully, you should be able to see the relative amount of iron metal in a cut slab of an ordinary chondrite and if you are well practiced you can possibly distinguish between the three groups by simply noting the amount of metal flakes in the matrix of a cut slice. The olivine composition, however, is another matter, as we saw, requiring an electron microprobe to analyze the composition of the olivine.

a

b

Figure 4.6. a Cook, an H chondrite with between 15% and 19% iron-nickel metal. **b** Beeler, an LL chondrite with less than 1% metal. Width of Cook 5.5 cm; width of Beeler 8.5 cm.

Table 4.1. The important classes and groups of chondrites are shown with their characteristic percentages of metal

Class	Group	Metal(wt%)	Total iron (wt%)	Fa (mole%)	Fs (mole%)
E	H&L	17–23	22–23	<1	0
OC	H	15–19	25–30	16–20	14–20
OC	L	1–10	20–23	21–25	20–30
OC	LL	1–3	19–22	26–32	32–40

Petrographic Types of Ordinary Chondrites

If we were to randomly cut slices from a dozen or so chondrites we would immediately notice that their interiors were packed with those spherical inclusions called chondrules. Chondrites show wide variations in chondrule density as well as definition from specimen to specimen. In some cases, the chondrules are so densely packed that little matrix material can be seen between them, even with a hand lens. Some chondrites show abundant well-defined chondrules while others seem to have faded. In some, the chondrules have almost disappeared into the matrix leaving only a hint of the chondrule fields that were once well-defined. What could have caused these modifications of the chondrules? Most researchers accept that chondrules formed early in the solar nebula's history and that the changes they see in chondrule texture is a result of secondary processing. Thermal metamorphism may have resulted from heat produced by the decay of radioactive aluminum called aluminum 26. This heating within the growing parent bodies recrystallized the matrix and chondrules causing a blurring of the chondrule textures. As the chondrules accreted onto growing planetesimals (asteroid parent bodies) they trapped heat from the accretion process. Stony silicate minerals are poor conductors of heat and as the parent bodies grew, they accreted layers of minerals that became hotter with increasing depth. Near the center of the parent body temperatures reached as much as 950 °C, not enough to melt the chondrules but sufficient to cause solid state recrystallization of the matrix and thermal metamorphism of the minerals as well. This variable chondrite texture has proven to be a useful tool in classifying chondritic meteorites.

In 1967, W. Randall Van Schmus and John A. Wood published an important paper that presented a comprehensive classification system for the chondrites which, with some modification, is still used today to classify chondrites into six petrographic types from type 1 to type 6. Originally, type 1 meteorites were thought to represent the lowest metamorphic grade but that distinction was later given to type 3. They used 10 criteria to define the petrographic types for all chondrites. We will briefly examine three of those that are readily observed with the unaided eye and/or microscope. A complete listing of the 10 criteria may be found in Appendix 2.

Perhaps the easiest of the 10 to study is chondrite texture. Table 4.2 shows a chart used to define seven different petrographic types. (Van Schmus and Wood listed only six types but researchers today recognize seven.) Note that increasing thermal metamorphism extends to the right from petrographic types 3–7. These types show progressive stages of thermal metamorphism up to the limiting temperature of 950 °C, well within the range of solid state recrystallization. Type 3 is considered to be chemically unequilibrated and therefore the most primitive, since it shows wide variations in the chemistry of its olivines and pyroxenes. On the other hand, types 4 to 7 are equilibrated. Their chemistry is more homogeneous than type 3.

Figure 4.7 shows thin section images of three ordinary chondrites with petrographic type 3, type 5 and type 6. All three were photographed at the same scale. You can see the effect of various grades of thermal metamorphism on the chondrule textures. In (a) type 3 we see well-defined and densely packed subspherical to spherical chondrules set within a fine-grained black matrix. The matrix is composed of the same minerals as the chondrules, only the crystals are microscopic in size making the matrix opaque. This meteorite is designated an unequilibrated L3 ordinary chondrite. Once again, the L is a chemical designation and stands for low metal; the number 3 designates the petrographic type, in this case a type 3. In (b) we see a type 5, an H5 ordinary chondrite showing considerable solid state recrystallization with fewer chondrules. The textures of the matrix and chondrules are less coarse than in (a). In type 6 (c), there are only a few chondrules remaining. Solid state recrystallization between matrix and chondrules blurs the chondrule boundaries and destroys the original meteorite texture.

In the above comparison, we considered type 3 meteorites to be unequilibrated. This simply means that type 3 meteorites have been relatively unaffected by thermal metamorphism, with

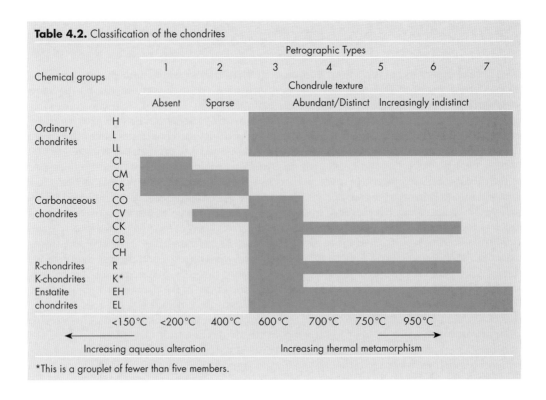

Table 4.2. Classification of the chondrites

Chemical groups		Petrographic Types						
		1	2	3	4	5	6	7
				Chondrule texture				
		Absent	Sparse	Abundant/Distinct		Increasingly indistinct		
Ordinary chondrites	H							
	L							
	LL							
	CI							
	CM							
	CR							
Carbonaceous chondrites	CO							
	CV							
	CK							
	CB							
	CH							
R-chondrites	R							
K-chondrites	K*							
Enstatite chondrites	EH							
	EL							
		<150 °C　<200 °C　400 °C　600 °C　700 °C　750 °C　950 °C						
		←　Increasing aqueous alteration			Increasing thermal metamorphism　→				

*This is a grouplet of fewer than five members.

matrices least recrystallized of all the chondrites. They are among the most primitive of all the meteorite groups and remain today in a relatively pristine state much as they were in the solar nebula 4.5 billion years ago. It is no wonder that type 3 chondrites are scientifically the most valuable of the chondrites. They tell us much about the processes in operation in the early Solar System. Collectors and researchers revere type 3 meteorites for much the same reasons. (An additional incentive to acquire such primitive specimens is that the more pristine and primitive they are, the more monetary value is placed upon them.) Type 3 ordinary chondrites are relatively rare. Unlike type 3 meteorites, which are unequilibrated, types 4–7 are considered equilibrated which means that these types have completed their chemical reactions with adjacent minerals and glass and have ceased to react. They have reached a state of chemical equilibrium. Their minerals have homogenized.

Matrix texture is another criterion that is useful in determining the petrographic type of the meteorite in question. We note that the matrix remains opaque through type 3 and into early type 4. At that point, we begin to see a transparent microcrystalline matrix forming. It continues to recrystallize, growing larger and larger crystals until it reaches type 6. At this point the matrix is transparent and light gray in color.

A final criterion that is a bit more challenging to the beginner at the microscope is the degree of development of secondary feldspar. Feldspar is a sodium-calcium aluminum silicate that begins to crystallize out of the matrix in submicrocrystalline grains in petrographic type 4. By type 5 the existing matrix glass begins to crystallize and disappear. Remaining crystals continue to increase in size in the matrix in type 6. The appearance of feldspar crystals of 50–100 μm in size is a definitive sign of an equilibrated type 6 matrix. Any further heating beyond that point and the chondrite may suffer melting.

a

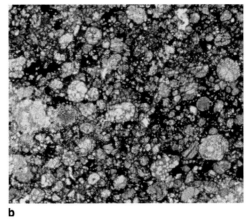

b

Figure 4.7. Thin section images of three ordinary chondrites are shown with petrographic type numbers indicating the textures of the stones. These are **a** type 3 (Moorabie, L3.6), **b** type 5 (Faith, H5) and **c** type 6 (Mbale, L6). Notice the clear distinction between matrix (black) and chondrules in type 3. In type 6, heating and recrystallization have destroyed the texture of type 3. All three images were made to the same scale.

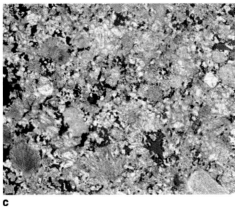

c

Ordinary Chondrites

> **Class: Ordinary Chondrites (OC)**
> Petrographic and chemical classifications: H, L, LL; 3–7
> Fall and find statistics: Falls—more than 739; Finds—more than 13,526
> Well-known examples: Holbrook, Mbale, NWA 869, Park Forest, Peekskill, Plainview

Today we have a classification system that takes into account both the chemistry and mineralogy of a meteorite as well as its texture modified by recrystallization. Since Van Schmus and Wood's original work on meteorite classification there have been some modifications. For example, many researchers now accept a type 7 classification for all three groups of the ordinary chondrites (H7, L7, LL7). Type 7 may be defined as those meteorites in which the chondrules have experienced partial (or complete) melting. This applies primarily to ordinary chondrites, however.

Notice the temperatures indicated on the bottom of Table 4.2. Increasing thermal metamorphism extends to the right with temperatures from 600 to 950 °C. Type 3 chondrites mark a dividing line between thermal metamorphism and a different type of processing—aqueous alteration. Some of the carbonaceous chondrites, primarily CI1, CM2, and CR2, have apparently suffered periods of severe alteration in which the meteorites reacted with water at room temperature or less (~20 °C). Most of these reactions probably took place on or within the asteroid parent body.

Ordinary chondrites are shown in Figures 4.8–4.15.

Figure 4.8. An assortment of Holbrook L6 stony meteorites, some of the more than 14,000 stones weighing totally about 218 kg that fell near the railroad yard east of Holbrook, Arizona in 1912. Individuals weigh between 6.6 kg to tiny specimens of only a few milligrams.

Figure 4.9. This individual stony meteorite was found adjacent to a cultivated field on May 11, 2006, near the town of Kackley, Kansas. It is a fully encrusted single stone weighing 1.36 kg. It was classified as an H4. Courtesy of Mark Bostick, Kansas Meteorite Society.

Figure 4.10. Ghubara, an L5, was found near Jiddat al Harasis, Oman in 1954. Since its initial discovery, hundreds of pieces have been found totaling over 500 kg. It is a xenolithic breccia composed of rock fragments not related to each other. The rock fragments in this example are indistinct and appear as patchy regions. Longest dimension 10 cm.

Figure 4.11. Plainview, an H5, is a regolith breccia with a mottled light/dark structure emphasized by weathering. It contains xenolithic rock fragments including carbonaceous and refractory inclusions (CAIs). Width 6 cm.

Figure 4.12. NWA 869 L4-6, brecciated chondrite, found in 2000, in a large strewn field. At least two metric tons have been recovered. Individual masses range from less than 1 g to more than 20 kg. It is one of the most readily available stony meteorites.

Figure 4.13. Golden Rule L5 found in 1999 by Twink Monrad while searching for Gold Basin specimens. Total weight: 798 g. Courtesy of Chris Monrad.

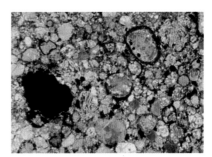

Figure 4.14. Portales Valley, a highly fractured H6 fell in 1998. Intrusion of metal wedges into silicate fractures within the matrix. Many angular silicate fragments display obvious signs of displacement. Photograph by Geoffrey Notkin/Aerolite. org, © The Oscar E. Monnig Meteorite Gallery.

Figure 4.15. Parnallee LL3.6 thin section in plane-polarized light shows typical chondrule field. Spherical chondrules are packed tightly in a dark matrix of similar composition. Large black object is an iron-nickel inclusion measuring 5 mm in longest dimension.

Group: Enstatite chondrites (E)
EH chondrites: Falls—8; Finds—117
Well-known examples: Abee, Itqiy, Sahara 97103
EL chondrites: Falls—7; Finds—31
Well-known example: DaG 734

E chondrites are relatively rare in meteorite collections and represent only about 2% of all stony meteorites. Only about 200 are currently known. They are mostly EH3, EH4 and EL6. Of considerable interest is a 3-mg mass discovered in a soil sample collected by Apollo 15 astronauts near Hadley Rille on the Moon. Though very small, this tiny specimen is still included in the total E chondrites listed in the Catalogue of Meteorites. E chondrites formed in an oxygen-depleted environment. Most of their iron occurs either as metal or as iron sulfide. Their pyroxene contains virtually no iron, only nearly pure magnesium silicate (enstatite) with less than 1 mole% iron-bearing fayalite. They are therefore called enstatite chondrites. Like the ordinary chondrites, E chondrites are subdivided into H and L groups depending upon their total iron. EH chondrites have about 30% total iron and more metal. EL chondrites have about 25% total iron and less metal. The easiest way to tell the difference between the EL and EH chondrites is to view them in thin section and note the size and shape of their chondrules. EL3 chondrules are well-defined and average about 550 µm in diameter. EH3 chondrules average around 220 µm in diameter and they are less well-defined.

E chondrites are shown in Figures 4.16–4.18.

Figure 4.16. This is a slice of an EH3 from Northwest Africa. Tilting the specimen in diffuse light will show that it is full of tiny metal particles, perhaps 35%. Iron in the pyroxene is depleted and therefore only the pure magnesium-rich enstatite remains. The chondrules, easily visible in this photo, are composed of nearly pure enstatite. Slice measures 23 mm.

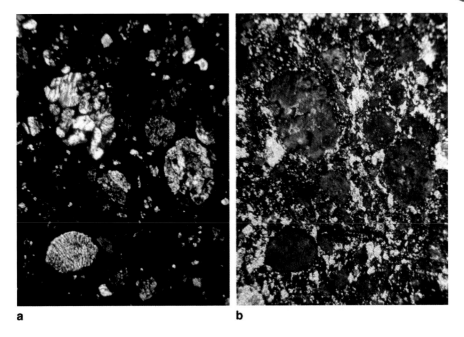

a   b

Figure 4.17. Seen in thin section, enstatite in this Sahara EH3 is abundant. In (**a**), the ubiquitous light gray interference colors are those of enstatite seen in cross-polarized light. Small rare spots of bright blue and green are olivine. In (**b**), metal (iron-nickel) shines brightly in reflected light. Typical proportions of the components of EH3 meteorites are 65% enstatite, 15–20% chondrules, 15% metal, 2–15% matrix, and less than 1% olivine.

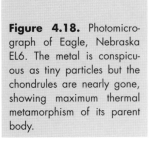

Figure 4.18. Photomicrograph of Eagle, Nebraska EL6. The metal is conspicuous as tiny particles but the chondrules are nearly gone, showing maximum thermal metamorphism of its parent body.

Group: R Chondrites (R), Rumurutiites
Type specimen: Rumuruti, fell in the Rift Valley, Kenya, 1934
Petrographic type: 3–6; Shock stage: S2
Total recovered weight of Rumuruti: 67 g

Fall and find statistics: Falls—1; Finds—78
Well-known example: Carlisle Lakes

In 1977, a badly weathered 49.5 g meteorite was found near Carlisle Lakes on the northern edge of the Nullarbor Plain in Western Australia. This meteorite and subsequent finds were so badly weathered that their most important distinguishing features could not be clearly determined. Forty years earlier (1934) a witnessed meteorite fall occurred near Rumuruti in southwestern Kenya. A number of stones were collected almost immediately afterwards. One specimen, a 67 g individual, was purchased by the Berlin Museum of Natural History. There it remained in a drawer unstudied until 1993 when its true identity was revealed. Very little weathering had occurred in the 40 years so important characteristics could be easily studied. This turned out to be a new meteorite group that could be matched with the earlier-found Carlisle Lakes meteorite. The single specimen at the Berlin Museum was given the distinction of becoming the type specimen for a new meteorite group called Rumuruti-like, Rumurutiites or R chondrites.

Recently many R chondrites have been discovered and recognized as meteoritic, largely from Antarctica and the hot deserts. With the exception of the Carlisle Lakes specimens, all are brecciated. Light-colored equilibrated clasts show a petrographic type 5–6 set against a dark fine-grained matrix of petrographic type 3–4. R chondrites are the most oxidized of all the chondrites. They are essentially free of iron metal, as opposed to the E chondrites which have almost all of their iron in the metallic state.

R chondrites are shown in Figures 4.19–4.22.

Figure 4.19. NWA 2921, an R3.8. An individual R chondrite (Rumurutiite) found in the Sahara in 2005. Individual shows remnant fusion crust covered by desert varnish. Total weight before slicing: 44.4 g. Courtesy of Jeff Kuyken, www.meteorites.com.au.

Figure 4.20. NWA 2921, an R3.8. After slicing, interior shows brecciation and numerous chondrules set in a fine-grained matrix. Courtesy of Jeff Kuyken, www.meteorites.com.au.

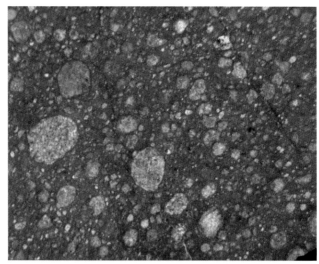

Figure 4.21. Thin section of NWA 753 R3.9 chondrite in plane polarized light showing large millimeter-sized chondrules and a sea of smaller chondrules and chondrule fragments. No free metal. Medium gray matrix similar to Allende CV3.

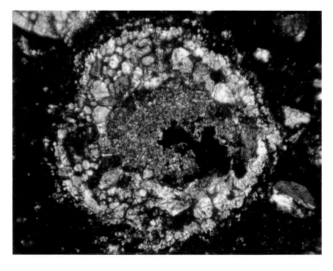

Figure 4.22. Thin section of NWA 753 R3.9 showing a large chondrule with a thick rim of bright olivine crystals and a cryptocrystalline olivine core surrounding an opaque interior of iron sulfide. Typically, R-chondrites are highly oxidized and therefore have very little metal. Seen here in cross-polarized light.

Carbonaceous Chondrites

We have spent considerable time with the basic fundamentals examining the most common ordinary chondrites. Now we can use what we learned from the ordinary chondrites to look at the most impressive stony meteorites—the carbonaceous or C chondrites.

Meteoriticists have developed a shorthand notation that serves to identify the different groups of carbonaceous chondrites. For example, for a CV3 meteorite: the "C" stands for the carbonaceous class, V is for the Vigarano group, and 3 is for the petrographic type. There are eight groups of C chondrites: CI, CM, CV, CO, CR, CK, CB and CH. There are also ungrouped meteorites that don't fit neatly into any of the established groups, and related groups that form clans. Some of the ungrouped specimens are totally unique. Among them are the remarkable Tagish Lake meteorites that fell in the Yukon Territory, Canada on January 18, 2000 (Figure 4.23).

It is estimated that the main body measured 4 m in diameter and weighed 56 metric tons at the top of the atmosphere. Fortunately, many of the meteorites fell onto Tagish Lake which was frozen at the time and were thus protected from abrupt aqueous alteration. About 97% of the meteoroid burned up in the atmosphere on the way down. It fragmented several times during its atmospheric flight that finally terminated some 28 km above the ground and the lake. Most of the surviving meteorites fell onto the Taku arm of the lake. It is estimated that hundreds of meteorites may have been lost in the lake and surrounding forest. The largest single fragment collected during the initial search weighed 159 g.

Tagish Lake meteorites turned out to be no ordinary C chondrites. Petrographically, they resembled CM chondrites with sparsely distributed small chondrules, altered CAIs (calcium-aluminum inclusions) and individual olivine grains not present in CI chondrites. The mineral composition was more like the most primitive carbonaceous chondrites, the CI chondrites. The high bulk carbon was more typical of the CI chondrites but other minerals were more typical of CM chondrites. It was finally decided that the classification should be intermediate between CI and CM, namely CI2.

Figure 4.23. These fragile pieces of the Tagish Lake CI2 meteorite managed to survive the fall through Earth's atmosphere. White grains of calcium aluminum inclusions (CAIs) are surrounded by an abundant matrix. Two or three "pseudomorphed" chondrules can be seen (circular white areas). Total weight: 262 mg.

Meteorites in the C groups are chemically some of the most complex, heterogeneous meteorites known. On the exterior, these meteorites remind us of charcoal briquets. Their fusion crusts are dark gray to black and their interiors are equally dark. Some show a plethora of well-shaped chondrules while others have nearly featureless interiors. Unlike the ordinary chondrites they show little if any sign of thermal metamorphism. Their compositions are very nearly like that of the Sun (minus the volatile elements). Many carbonaceous chondrites are further distinguished from the ordinary chondrites by their lack of metal. What metal they may have acquired in their formative early years (4.56 billion years ago) has been combined with oxygen, forming oxides like magnetite. Perhaps the single most important characteristic is the presence of water-bearing minerals. In some carbonaceous chondrites there is evidence that liquid water has percolated through these meteorites via fractures. This water reacted chemically with the original minerals (olivine and pyroxene) at temperatures well above the freezing point. This created hydrated silicate minerals similar to those found in terrestrial clays and serpentine. Unlike the ordinary chondrites, virtually all carbonaceous chondrites show signs of aqueous alteration. It seems to be a hallmark of many of the carbonaceous chondrite groups.

When the carbonaceous chondrites were first studied, many researchers accepted the notion that they contained substantial amounts of carbon, more carbon than in other chondrites. This turned out not to be true. There are some carbonaceous chondrites that are actually carbon-poor. The abundance of carbon is not the main characteristic of these meteorites, however. Rather it is their higher abundance of magnesium, calcium, and aluminum relative to silicon than that found in ordinary chondrites. Some of the most primitive carbonaceous chondrites (CI1) contain carbonates and complex organic compounds such as amino acids that could be involved in the origin of life. Let us now look at the individual meteorite groups that define the carbonaceous chondrites.

Group: CI1 Carbonaceous chondrites (CI)
Type specimen: Ivuna, fell in Mbeya, Tanzania, 1938
Shock stage: S1
Total recovered weight of Ivuna: 705 g

Fall and find statistics: Falls—5; Finds—2
Well-known examples: Alais, Orgueil, Revelstoke, Tonk

There are only 7 CI carbonaceous chondrites known in the world. Most were observed falls. Of these, only two deposited sufficient material for any meaningful scientific research. The type specimen, Ivuna, fell to Earth near the town of Ivuna in Tanzania on December 16, 1938. (Meteorites are usually named for a nearby geographic location or town.) Carbonaceous chondrites follow the petrographic and chemical classification scheme of the ordinary chondrites. Earlier in this chapter we noted that ordinary chondrites were given a petrographic designation between 3 and 7, 3 being the least thermally altered and 7 the most thermally altered. This is determined by the texture of the interior, which in turn is a function of the thermal metamorphism under which the meteorite originally formed. Carbonaceous chondrites are classified much the same way but these meteorites have not been altered by heat. Instead, they have been altered through the action of water. Ivuna has been classified a petrographic type 1. The matrix is fine-grained and opaque. The carbon content is 3–5% and the bulk water content is between 18 and 22%. These three criteria alone are sufficient to classify these chondrites as a type 1. The strongest and most surprising criterion is the chondrule texture—there are no chondrules. Orgueil is the most abundant of the five CI chondrites in that over 20 individuals totaling 14,000 g were collected. Today, most of the scientific work on CI1 chondrites, especially as it relates to the origin of life, has been through studies of Orgueil. It is a much sought-after meteorite and an extraordinary prize for the collector.

CI1 chondrites are shown in Figures 4.24–4.26.

Figure 4.24. Ivuna, one of the most famous meteorites in the world, is the type specimen for the CI carbonaceous chondrites. The dark interior of carbonaceous meteorites is well illustrated in this photo. This piece weighs 17 g. Courtesy of Luc Labenne, Labenne Meteorites, www.meteorites.tv.

Figure 4.25. One of the 20 stones recovered from the fall of Orgueil, a CI carbonaceous chondrite. This piece weighs 4 g. Courtesy of Luc Labenne, Labenne Meteorites, www.meteorites.tv.

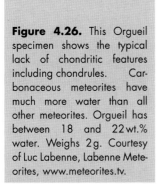

Figure 4.26. This Orgueil specimen shows the typical lack of chondritic features including chondrules. Carbonaceous meteorites have much more water than all other meteorites. Orgueil has between 18 and 22 wt.% water. Weighs 2 g. Courtesy of Luc Labenne, Labenne Meteorites, www.meteorites.tv.

Group: CM chondrites (CM)
Type specimen: Mighei CM2, fell in Nikolayev, Ukraine, 1889
Shock stage: S1
Total recovered weight of Mighei: 8 kg

Fall and find statistics: Falls—15; Finds—146+
Well-known examples: Adelaide, Cold Bokkeveld, Murchison, Murray

Today when we think of CM2 carbonaceous chondrites we think of the fall of Murchison. On September 28, 1969, an extraordinary fall of over 700 stony meteorites rained down upon the town of Murchison, Victoria, Australia. They fell in front yards, on roof tops and in streets. Over 100 kg of pristine, rare carbonaceous stones, some still warm to the touch, were recovered and taken indoors by the inhabitants. Many of the stones displayed striking flow structures. Others had oriented cone shapes. CM chondrites are all type 2 and are the first carbonaceous meteorite types in which chondrules appear, well-formed but small and sparsely distributed in the black, opaque matrix. Olivine is abundant both as grains and small chondrules. Mighei, the prototype CM2 specimen, has a chondrule abundance of about 20 vol.% and a chondrule mean diameter of 0.3 mm. The chondrules and other inclusions are set in a fine-grained black, opaque matrix. Major constituents of the matrix are hydrated phyllosilicates similar to terrestrial clays. These meteorites show a moderate degree of aqueous alteration (chemically reacting with water). They contain between 3 and 11 wt.% water. Murchison contains between 20.44% and 22.13% total iron. (Total iron includes both chemically combined and elemental iron.) Grains of metal (kamacite) and iron sulfide (troilite) appear in the matrix and within olivine aggregates totaling between 4 and 11 wt.%. Most exciting since the fall of Murchison has been the discovery within the meteorite of organic compounds: namely amino acids, the building blocks of proteins, including left-handed amino acids, the type found in all living things.

CM2 carbonaceous chondrites are shown in Figures 4.27–4.30.

Figure 4.27. This is Mighei, the type specimen for the CM2 carbonaceous chondrites. Small light gray CAIs and small chondrules can be seen in the dark matrix. Serpentine, a product of hydrous alteration, gives a green tint to the matrix. This piece weighs 4.8 g. Photograph by Iris Langheinrich, R. A. Langheinrich Meteorites, www.nyrockman.com.

Figure 4.28. The best known CM2 and one in which a great deal of research has been done is Murchison. It is noted for the beauty of its fusion crust. This stone shows flow lines marking rivulets of molten silicates that streamed along its structure. 107 g. Courtesy of Jim Strope, www.catchafallingstar.com.

Figure 4.29. This fragment of Murchison shows its smooth exterior and fine-grained black matrix. Even in thin section, the matrix of this and all other carbonaceous chondrites are black and opaque. The subspherical structures are chondrules. Specimen weighs 42 g.

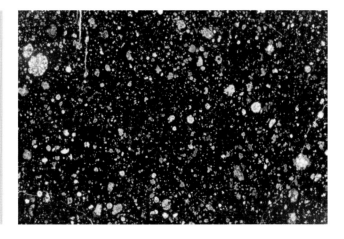

Figure 4.30. This thin section of Murchison shows a sampling of small porphyritic olivine chondrules and pieces of chondrules scattered in the opaque matrix. Fine fractures permeate the interior. Plane-polarized light.

Group: CV Carbonaceous chondrites (CV)
Type specimen: Vigarano CV3.3, fell in Emilia-Romagna, Italy in 1910
Shock stage: S1–S2
Total recovered weight of Vigarano: 15 kg

Fall and find statistics: Falls—7; Finds—106
Well-known examples: Allende, Axtell

Before 1969, the total number of CV chondrites stood at about 97 specimens. They were some of the rarest of the carbonaceous chondrites. Then, on the night of February 8, 1969, at 1:05 a.m. over two tons of CV3.2 chondrites fell over Chihuahua, Mexico, near the little town of Pueblito de Allende. It remains, numerically, one of the largest multiple falls ever witnessed. The fall of Allende made more CV chondrites available for study and distribution than had existed altogether in the world before. Today, Allende is one of the most studied CV chondrites. It makes an excellent display specimen for museums and private collectors, and generally at a reasonable price.

CV exteriors are very different from those of either the CI or CM2 chondrites. CI and CM2 crusts are opaque and black, making it difficult to distinguish between the fusion crust and interior. Allende's fusion crust is a dark gray but encloses a lighter medium gray interior. Breaks in the crust reveal a large abundance of chondrules averaging 1 mm in diameter. Allende's chondrules are among the largest found in the carbonaceous chondrite groups. Would you believe 25 mm is the record? Type CO chondrules, by contrast, are quite small, averaging about 0.15 mm. CV chondrule abundances average between 35 and 45 vol.%. Only the ordinary chondrites have larger chondrule fields (60–80 vol.%).

Studying a large slab of Allende we come across structures normally not seen or seen only rarely in other chondrites. These are the highly refractory minerals which formed in the solar nebula several million years before the first chondrules formed. They are called CAIs or calcium-aluminum inclusions and are found prominently in all CV3 chondrites. Allende contains between 5 and 10 vol.% CAIs. The mineralogy of CAIs is complex, being composed of highly refractory oxides such as melilite (calcium aluminum oxide), spinel (a magnesium aluminum oxide) and anorthite (a calcium aluminum silicate), to name a few. Among these minerals, especially the anorthite, excess ^{26}Mg is found in crystal lattices suggesting that the isotope ^{26}Al at one time existed in those locations and decayed to stable ^{26}Mg. Thus, ^{26}Al existed as a precursor isotope in the early solar nebula.

CV chondrites are shown in Figures 4.31–4.35.

Figure 4.31. The black primary crust of this CV3.2 carbonaceous chondrite from the Allende fall contrasts sharply with the medium gray interior. Note the regmaglypts (thumb prints). This stone weighs 1,707 g. Photograph by Geoffrey Notkin/Aerolite. org, © The Michael Farmer Collection/www. meteoritehunter.com

Figure 4.32. A slab of Allende CV3.2 which is 14cm wide. About half of the inclusions seen here are chondrules. The lightest colored inclusions are CAIs. The dark elongated object in the center is probably a fragment of a CO carbonaceous chondrite with tiny chondrules. Weighs 317g.

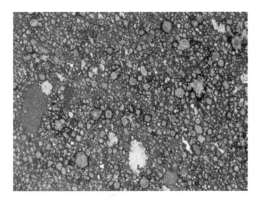

Figure 4.33. A slice of Axtell, a CV3.0 found in Texas in 1943. Here, as in all CV chondrites, chondrules make up ~45%, matrix ~40%, CAIs ~10%, and metal 0–5%. This piece weighs 264g and measures 9 × 12cm. Photograph by Geoffrey Notkin/Aerolite.org, © The Oscar E. Monnig Meteorite Gallery.

Figure 4.34. A detail of the slice of Axtell. Notable are large numbers of chondrules, subangular gray inclusions (xenoliths) of many sizes, and white irregular-shaped CAIs. The matrix shows pervasive weathering (weathering grade 5) seen here as a medium brown. Photograph by Geoffrey Notkin/Aerolite.org, © The Oscar E. Monnig Meteorite Gallery.

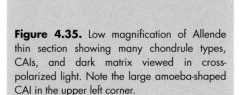

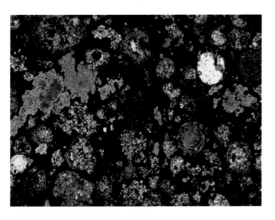

Figure 4.35. Low magnification of Allende thin section showing many chondrule types, CAIs, and dark matrix viewed in cross-polarized light. Note the large amoeba-shaped CAI in the upper left corner.

Group: CO Carbonaceous chondrites (CO)
Type specimen: Ornans CO3.4, fell in Franche-Comte, France 1868
Shock stage: S1
Total recovered weight of Ornans: 6 kg

Fall and find statistics: Falls—6; Finds—153
Well-known examples: Kainsaz, Lancé, Moss

Like most of the members of the carbonaceous chondrite class, the chondrites in this group are named after the type specimen. The Ornans meteorite is related both physically and chemically to the CV and CK chondrites which together form a distinct clan. The most outstanding difference between the CV chondrites and the CO chondrites is in the comparative size of their chondrules. CV chondrules average about 1 mm in diameter. The CO chondrites are loaded with tiny chondrules usually less than 0.2-mm diameter. They are so small and densely packed in the matrix that they make up over 48 vol.% of the meteorite. Like the CV chondrites, CO chondrites have refractory inclusions (CAIs) covering about 15% of the matrix, but these inclusions are usually much smaller and more sparsely distributed within the matrix than are the CV chondrites. Free metal (FeNi) in small inclusions are frequently seen in well-prepared polished slabs.

On July 14, 2006, a new CO3.5/3.6 fell in Norway. It has been named Moss. Five stones, with a total known weight of 3.76 kg, struck a roof and a fence. The pieces were spread along a line over 4 mi (6 km) in length.

CO chondrites are shown in Figures 4.36–4.38.

Figure 4.36. A piece of Ornans CO3.3, the type specimen for the CO carbonaceous chondrites, without crust. It has tiny chondrules (0.15 mm) and nearly 35 vol.% matrix. University of New Mexico collection.

Figure 4.37. a A freshly fallen CO3.5/3.6 specimen from Moss that fell in Norway in July, 2006. It displays a fresh black fusion crust, a light-colored interior, a friable loose texture, and chondrules that are difficult to distinguish. **b** Polished slice showing bright metal in the interior. CO chondrites typically have up to 5% metal. Courtesy of Mike Farmer, www. meteoritehunter.com.

Figure 4.38. Dar al Gani 749. This image shows the complex texture of the cut face of a CO3 chondrite. Metal fields in the interior can be detected with a strong magnet. The orange areas are the result of terrestrial weathering. Width 30 mm.

Group: CK Carbonaceous chondrites (CK)
Type specimen: Karoonda, fell in South Australia, 1930
Shock stage: S1
Total recovered weight of Karoonda: 41.73 kg

Fall and find statistics: Falls—2; Finds—153
Well-known example: Maralinga

The CK chondrites are known almost entirely from specimens found in Antarctica, beginning around 1990 and increasing rapidly until 73 were known by the century's close. Of the 73, only two were witnessed falls. The first landed in Karoonda, South Australia in 1930. The other was seen to fall near the town of Maralinga on the Nullarbor Plain of South Australia. Maralinga is the most readily available of the CK chondrites. Most of the CK group are highly equilibrated, having petrographic types 4–6 with 75% equilibrated to type 5. They are the only carbonaceous chondrite group showing petrographic types beyond 3. A cut and polished specimen shows a dull, blackened interior field. This blackening is called silicate darkening. The darkening makes the field difficult to distinguish with a hand magnifier or microscope. Darkening is a trait noted in all CK chondrites and is the result of very fine-grained magnetite and pentlandite (a nickel-iron sulfide) permeating the silicate (olivine) interior. Chondrules cover about 45 vol.%, with each chondrule averaging about 1.0-mm diameter.

CK chondrites are highly oxidized. That they show no metal grains in their matrix along with the presence of magnetite, troilite, and iron-bearing silicates is a sure sign of the high degree of oxidation. Instead, their olivines and pyroxenes are iron-rich. There may be some fine-grained iron minerals in the matrix that could be detected with strong magnets. The chemistry of CK chondrules is similar to CO and CV chondrites but significant differences in their chemistry tend to set them apart from the CO and CV chondrites.

CK chondrites are shown in Figures 4.39–4.42.

Figure 4.39. Karoonda CK4 is the type specimen for the CK chondrites. Width 64 mm. Robert Haag collection.

Figure 4.40. A 28-mm slice of the Maralinga CK4 chondrite, which exhibits moderate weathering. The largest chondrule in the field is 1.5 mm.

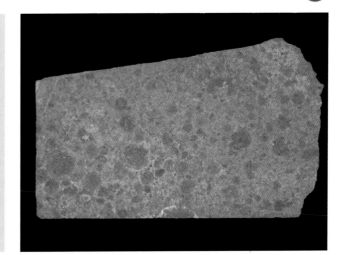

Figure 4.41. Thin section of Maralinga in plane-polarized light. Like other CK4s, it shows a dull gray silicate darkening due to fine-grained pentlandite and magnetite that have grown inside clear silicate minerals in the matrix. Orange and brown areas are weathered metal and sulfides.

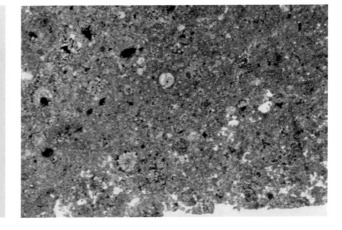

Figure 4.42. NWA 765, a CK4/5 found in Morocco in 2000. Dark chondrules are scattered throughout a sooty gray matrix. Courtesy of Bruno Fectay and Carine Bidaut, The Earth's Memory, meteorite.fr.

> ## Group: CR Carbonaceous chondrites (CR)
> Type specimen: Renazzo CR2, fell in Emilia-Romagna, Italy in 1824
> Shock stage: 1–3
> Total recovered weight of Renazzo: 10 kg
>
> Fall and find statistics: Falls—3; Finds—105
> Well-known examples: Acfer 059, NWA 801, Tafassasset

The CR carbonaceous chondrites are a relatively new group. It was not recognized as a group until similar specimens were recovered in the 1980s by scientists searching for meteorites in Antarctica and a decade later by collectors combing the Sahara Desert. Renazzo became the type specimen for the CR chondrites in the 1990s.

The CR chondrites differ from most groups of carbonaceous chondrites in that they contain substantial amounts of metal, usually associated with chondrules, i.e., with metal rims. The chondrule sizes average about 0.7 mm, and taken together with chondrule fragments make up as much as 50 vol.% of the meteorite. All CR chondrites are petrographic type 2, indicating that these chondrites show some aqueous alteration.

CR chondrites are shown in Figure 4.43–4.45.

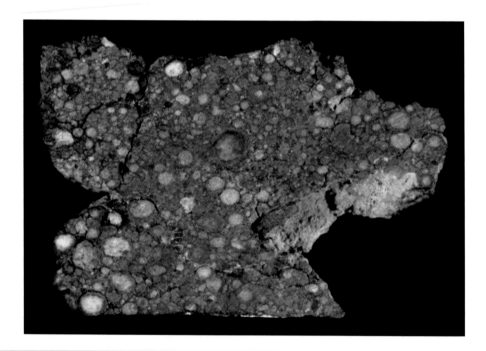

Figure 4.43. This is Acfer 059, a CR2 chondrite found in Algeria in 1989. It has similarities to Renazzo, the type specimen. The chondrules, with an average diameter of 0.7 mm, have conspicuous darkened rims due to hydrous alteration to phyllosilicates. This piece weighs 28 g and measures 6 cm. Robert Haag collection.

through a petrographic microscope. So stunning are these colors and shapes that we have included a Chondrule Gallery to illustrate these basic forms and more. Much as imaging distant celestial objects requires modern CCD electronics, the internal beauty of a stony meteorite requires specialized equipment to be revealed. Chap. 11 discusses this process in greater detail (Figure 4.49–4.72).

Table 4.3. Classification of chondrule types based on texture (After Keil and Gooding, 1981.)

	Type	Texture and Minerals	Abundance(%)
Group 1 (porphyritic)	PO	porphyritic olivine	23
	PP	porphyritic pyroxene	10
	POP	porphyritic olivine-pyroxene	48
Group 2 (nonporphyritic)	RP	radial pyroxene	7
	BO	barred olivine	4
	C	cryptocrystalline	5
Group 3	GOP	granular olivine-pyroxene	3

Figure 4.49. An exceptionally large chondrule 7 mm in diameter embedded in matrix. From an Allende CV3.2 carbonaceous chondrite.

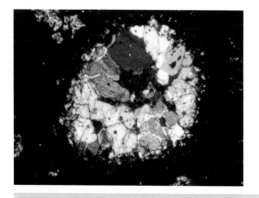

Figure 4.50. A porphyritic olivine (PO) chondrule from an Allende CV3.2 carbonaceous chondrite. Seen in cross-polarized light. Courtesy of Dr. Tom Toffoli.

Figure 4.51. A porphyritic pyroxene (PP) chondrule in cross-polarized light from a Zegdou H3 chondrite. Courtesy of Dr. Tom Toffoli.

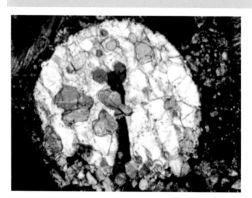

Figure 4.52. A porphyritic olivine pyroxene (POP) chondrule from the LL4 OC, Hamlet. Brightly colored grains of olivine are completely enclosed and surrounded by white orthopyroxene. This is called poikilitic texture. Courtesy of Dr. Tom Toffoli.

Figure 4.53. This chondrule from the Saratov L4 OC is a radial pyroxene (RP). The entire chondrule appears to be coated with colorful blocky clinopyroxene grains. The grains seems to elongate along a radial path leading to the core of the chondrule.

Figure 4.54. Fine-grained radial pyroxene (RP) chondrule from an NWA 520 L5 OC. Bundles of orthopyroxene (enstatite) fibers radiate from the side of the chondrule. These chondrules were probably completely melted when recrystallization commenced.

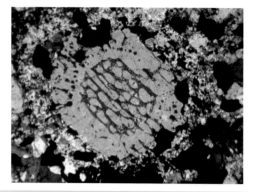

Figure 4.55. Barred olivine (BO) chondrule from L5 Beaver OC. Skeletal bars of olivine are separated by recrystallized glass with the entire chondrule surrounded by a thick igneous rim. The brilliant magenta color illustrates a high order of birefringence.

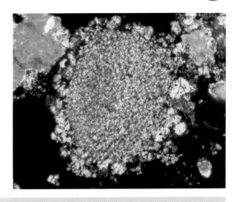

Figure 4.56. Cryptocrystalline (C) chondrule in a Sahara chondrite has a dark gray color typical of a mass of submicroscopic orthopyroxene crystals. Two chondrule "craters" are visible on lower left and mid right.

Figure 4.57. Granular olivine-pyroxene (GOP) chondrule in Allende CV3.2 carbonaceous chondrite contains a mixture of fine-grained equigranular olivine and pyroxene with grain sizes between 25 and 400μm in the core of the chondrule.

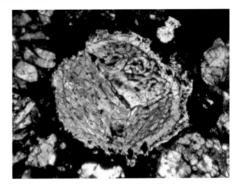

Figure 4.58. Radial pyroxene (RP) from Marlow L5 OC. This remarkable chondrule seems to be composed of at least three bundles of pyroxene fibers extending from the same point. Large compound "crater" is on lower right segment.

Figure 4.59. Chondrule from Moorabie L3.8 OC composed of three distinctive areas of olivine. In each area, the crystals of olivine have different orientations distinguished by their colors. A thick partially destroyed rim encircles the interior.

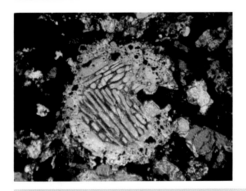

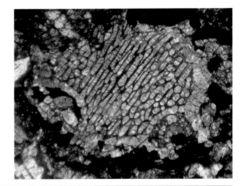

Figure 4.60. This remarkably beautiful chondrule from the Nuevo Mercurio H5 OC is a barred olivine (BO) growing in two directions with different crystal orientations. Recrystallized glass is interspersed between the olivine bars. A thick igneous rim, destroyed on the right side, supports the bars.

Figure 4.61. A curious chondrule from the Barratta L3.8 OC. It is one large crystal made of rows of short barred olivine with a rim of the same material. The chondrule has short projections, the remnants giving it a scalloped appearance.

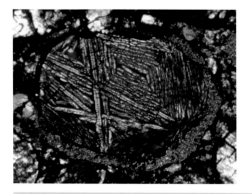

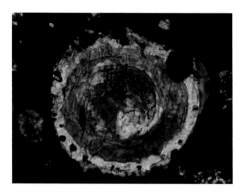

Figure 4.62. Complex chondrule from Marlow L5 OC shows several intersecting fields of thin olivine bars. Metal made of low nickel kamacite surrounds the chondrule as an incomplete metal jacket. This photo taken in cross-polarized and reflected light.

Figure 4.63. A striking multi-shelled chondrule of thick and thin layers of olivine. The alternating red and blue shells suggest either changing crystallographic orientations or alternating compositions of olivine. Moss, CO3.5. Courtesy of John Kashuba.

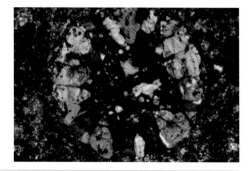

Figure 4.64. This very strange chondrule, resembling a "jellyfish," is in actuality half of a barred olivine chondrule in Barratta L3.8 OC. Fine threads appear to hang beneath the jellyfish's body, much as in real life. An S4 shock stage suggests microscopic shock features abound.

Figure 4.65. Maralinga CK4 carbonaceous chondrite. A circle of large subhedral olivine grains displays differing interference colors. The large grains appear to be radial to the chondrule center. Dark, fine-grained matrix material encloses the ring. Courtesy of Dr. Tom Toffoli.

Figure 4.66. Moorabie L3.6 OC. A chondrule, resembling a "horned owl," seen in cross-polarized transmitted light, reveals barred olivine with four crystal domains at right angles to each other, encircled by a thick rim. Large areas appear opaque suggesting intervening metal.

Figure 4.67. The same chondrule in reflected light plus cross-polarized light reveals metal in these spaces. The rim is permeated with tiny metal grains and much metal is scattered in the matrix. Some silicate grains are in extinction but appear here in reflected light.

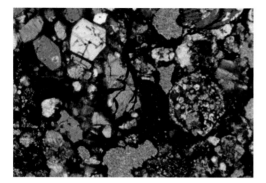

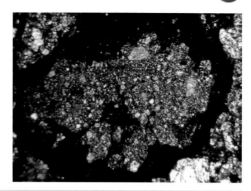

Figure 4.68. Cleo Springs H4 OC in cross-polarized light. Nearly perfect euhedral olivine crystals are set in a black matrix (above, left). A complete microchondrule filled with tiny olivine grains is in the middle right half of the photo.

Figure 4.69. Allende CV3.2 with numerous aggregates of tiny olivine grains (amoeboid olivine aggregate). It reminds us of an amoeba with tiny projections around the perimeter. The olivine grains appear almost identical in size (equigranular).

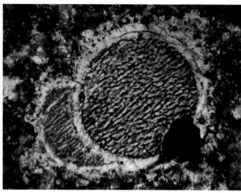

Figure 4.70. This wonderfully complex chondrule from the Barratta L3.8 OC is oddly elliptical in shape with a thin external rim enclosing several barred olivine fields. Two small compound chondrules appear in the dark matrix at the two and eight o'clock positions.

Figure 4.71. Perfectly symmetrical barred olivine chondrule from an unnamed Northwest Africa OC. Two chondrules, a primary and a secondary, make up a compound chondrule. Courtesy of Dr. Tom Toffoli.

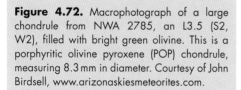

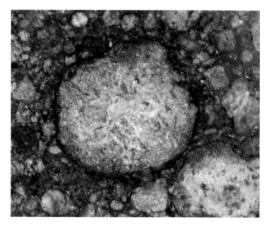

Figure 4.72. Macrophotograph of a large chondrule from NWA 2785, an L3.5 (S2, W2), filled with bright green olivine. This is a porphyritic olivine pyroxene (POP) chondrule, measuring 8.3 mm in diameter. Courtesy of John Birdsell, www.arizonaskiesmeteorites.com.

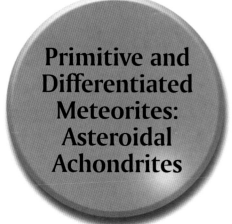

CHAPTER FIVE

Primitive and Differentiated Meteorites: Asteroidal Achondrites

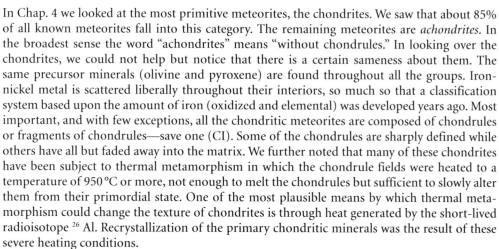

In Chap. 4 we looked at the most primitive meteorites, the chondrites. We saw that about 85% of all known meteorites fall into this category. The remaining meteorites are *achondrites*. In the broadest sense the word "achondrites" means "without chondrules." In looking over the chondrites, we could not help but notice that there is a certain sameness about them. The same precursor minerals (olivine and pyroxene) are found throughout all the groups. Iron-nickel metal is scattered liberally throughout their interiors, so much so that a classification system based upon the amount of iron (oxidized and elemental) was developed years ago. Most important, and with few exceptions, all the chondritic meteorites are composed of chondrules or fragments of chondrules—save one (CI). Some of the chondrules are sharply defined while others have all but faded away into the matrix. We further noted that many of these chondrites have been subject to thermal metamorphism in which the chondrule fields were heated to a temperature of 950 °C or more, not enough to melt the chondrules but sufficient to slowly alter them from their primordial state. One of the most plausible means by which thermal metamorphism could change the texture of chondrites is through heat generated by the short-lived radioisotope ^{26}Al. Recrystallization of the primary chondritic minerals was the result of these severe heating conditions.

On the left side of the temperature gradient in Table 4.2, beginning at about 400 °C and extending below about 150 °C (some as low as 20 °C), we noticed alteration due to the presence of water at some time in the past. This is especially noticeable in type 1 and type 2 chondrites. Here, fluids have entered the rocks through tiny fractures, subsequently reacting with the original primary minerals and producing hydrated minerals such as magnetite and clay-like phyllosilicates. The least affected chondrites are those of type 3 wherein there is scarcely any sign of aqueous alteration or thermal metamorphism that might have changed the primary characteristics of the meteorite.

Fifteen percent of the remaining meteorites are achondrites, irons and stony-irons. These three important meteorite types have little in common with the chondrites, but they have much in common with each other. Unlike the chondrites, achondrites formed by melting deep within their asteroid parent bodies. They were once chondritic but that primary structure was destroyed during planet building.

Differentiation

Very early in the Solar System's history Earth and the inner planets shared a more or less uniform composition much like the chondritic parent bodies. Their masses grew steadily as they accumulated rocky debris scattered throughout the inner Solar System. The forming planets heated gradually as the nearly continuous impact of rocky debris released heat, melting surface and near-surface rocks. This additional mass compressed the forming planets, releasing gravitational energy and converting it to heat deep in their interiors. Adding to this heat was the heat of decay of radioisotopes trapped in the deepest rocks. All these heat sources added together provided sufficient heat energy to melt the young, forming planets entirely. Now in a semi-liquid to liquid state, the planets began to differentiate (Figure 5.1). The formation of an iron core began the process in which the terrestrial planets gradually became layered. Under molten and/or semi-molten conditions the more dense minerals separated from the less dense. Heavy elements such as iron, nickel and some of the noble metals, gold, platinum and iridium separated from the forming viscous fluids, gradually sinking to their centers, forming heavy cores. Lighter elements and minerals accumulated around the core, forming dense basaltic rock and a thick mantle of lighter minerals. Differentiation was completed when the lightest minerals such as feldspars, quartz and mica slowly floated to the top, forming a relatively thin outer crust. The achondritic meteorites probably represent the outer crust of these differentiated parent bodies.

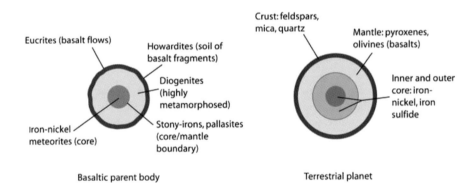

Figure 5.1. The basaltic achondrites came from parent bodies that became hot enough to completely melt. When melted, the processes of differentiation began and resulted in the segregation of liquids and crystals into separate concentric regions. The much larger terrestrial planets completely melted leaving no trace of their original building blocks.

Achondrites

The achondrites are the largest class of differentiated meteorites and include meteorites from the asteroid belt, the Moon and the planet Mars (Figure 5.2). In addition, there are very rare meteorites, the *primitive achondrites*, whose members show signs of partial melting and partial differentiation. The primitive achondrites are divided into three subgroups: acapulcoites, lodranites, and winonaites. They have important similarities and may have originated on the same asteroid parent body.

The primitive achondrites were only partially melted, but the *basaltic* (magmatic) achondrites are the product of complete melting and they are similar to terrestrial basalt. There are three groups of basaltic achondrites—eucrites, diogenites, and howardites (the HED association), which represent a collection of both volcanic and plutonic rocks formed from basaltic magmas. They are usually studied together as components of asteroidal basalt. HED meteorites are thought to have come from the same parent body, possibly from the asteroid 4 Vesta. Because of Vesta's unusual basaltic composition and especially the good match between Vesta's spectra and that of the HED meteorites, the relationship is thought to be well-established. The HED meteorites are the largest group of asteroidal achondrites with over 751 finds and falls combined.

Four additional rare groups complete the asteroidal achondrites: angrites, aubrites, ureilites, and brachinites. Finally, some of the rarest and most valuable achondrites are the planetary achondrites. They include 34 Martian and 56 Lunar meteorites finds. We will look at them in Chap. 6.

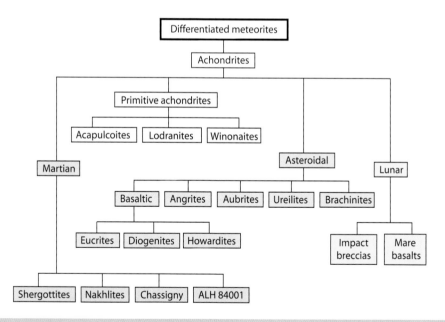

Figure 5.2. A flow chart showing the achondrites which include meteorites from the asteroid belt, Mars, the Moon and the primitive achondrites.

Primitive Achondrites

Acapulcoites (ACA); Lodranites (LOD); Winonaites (WIN)
Type specimen: Acapulco; fell in Guerrero, Mexico,1976
Total weight of Acapulco: 1,914 g

Fall and find statistics: Falls—1; Finds—39
Well-known specimens: Dhofar 125, Monument Draw

Type specimen: Lodran; fell east of Lodhran, Pakistan in 1868
Total weight of Lodran: 1,000 g

Fall and find statistics: Falls—1; Finds—17
Well-known specimens: NWA 2235

Type specimen: Winona, found in a stone burial cist in Arizona in 1928
Total weight of Winona: 25 kg

Fall and find statistics: Falls—1; Finds—20

Acapulcoites and lodranites belong to the small group of meteorites called primitive achondrites and share both chemical and physical characteristics. Because they are closely related both chemically and mineralogically, they are usually considered together. A fundamental distinction between the two primitive achondrites is in their comparative grain size. The acapulcoites have fine-grained chondritic abundances of orthopyroxene, the primary mineral. There is also olivine, FeNi metal, troilite, and chromite with grain sizes between 0.2 and 0.4 mm. Lodranites have grain sizes of 0.5–1.0 mm with a dominant silicate of olivine. Their composition is between the enstatite or E chondrites and H chondrites. Both the acapulcoite and lodranite meteorites are loaded with curiosities. For example, both have retained chondritic compositions and some acapulcoites actually bear chondrules. Lodranites have coarser-grained textures than acapulcoites but both are clearly achondritic.

Both meteorites show extensive recrystallization through thermal metamorphism, which is evident from the many 120 °C triple junctions appearing in thin sections. In thin section some differentiated meteorites show mineral grains of similar size crowded into one another. This is a good indicator that these meteorites came from a region within their parent body that was thermally metamorphosed. The buried region became hot for a long time but did not melt. Heat allowed the atoms of the original minerals to migrate short distances within the solid rock and rearrange themselves into a much simpler conFiguration of the same or sometimes different mineral. On Earth the full range of change due to thermal metamorphism can be studied. Rocks only slightly metamorphosed retain most of their original minerals and textures. But those that have been thoroughly metamorphosed retain few or none.

The coarser grains of the Lodranites suggest that they formed in deeper layers of their parent body where they were subjected to more intense thermal processing. Recently, some researchers have suggested that at least some of the primitive achondrites are actually residues from the partial melting of chondrites. They agree that these meteorites were in a process of differentiation (melting and segregation), beginning their conversion to achondrites. But the conversion was not completed. Thus, they remain in a transitional form between chondrites and achondrites. Both meteorites probably came from the same parent body, most probably an S-type asteroid.

Winonaites are named for the type specimen, which was found at Winona, Arizona in 1928, in a Native American stone burial cist at the Elden Pueblo. The site was occupied from 1070 to 1275 by the people now called Sinagua. The meteorite was buried in the same manner as children were buried in that culture, suggesting that it was highly revered, possibly because it was a witnessed fall. Although the meteorite was originally one mass, it fell apart when the cist was opened due to a long period of terrestrial weathering. There are 21 winonaites known, all fine to medium-grained, mostly equigranular and occasionally containing regions of relict chondrules. Mineralogically they are similar in composition to chondrites (intermediate between E and H). They contain iron, nickel and ferric sulfide veins which may represent the first partial melts on the original precursor parent body. They are closely related to the silicate inclusions which are found in IAB and IIICD irons. For this reason, Winonaites are often classed with the irons, but they are considered here with their achondritic cousins.

Primitive achondrites are shown in Figures 5.3–5.8.

Figure 5.3. This close-up photo of NWA 2714 shows the equigranular texture of a lodranite, which looks nearly identical to an acapulcoite. Multicolored crystals represent olivine, orthopyroxene, plagioclase, troilite, and kamacite. Crystal sizes are less than 1 mm. Field of view 25 mm. Courtesy of John Birdsell, www.arizonaskiesmeteorites.com.

Figure 5.4. Thin section of Monument Draw, an acapulcoite from Texas, shows bluish-gray metal and bronze-colored troilite (far right) scattered among equigranular gray orthopyroxene and colorful olivine crystals in this extensively metamorphosed meteorite. Crystal sizes are less than 0.5 mm. Cross-polarized and reflected light.

Figure 5.5. Thin section of acapulcoite NWA 2871, found in 2005, highlights some primary features of acapulcoites: colorful olivine and gray-to-white orthopyroxene, equigranular texture (grain sizes in a narrow range), and 120-degree junctions where three crystals meet. Cross-polarized light. Courtesy of John Kashuba.

Figure 5.6. NWA 2235 is a primitive achondrite, a lodranite, found in 2000. It is a coarse-grained aggregate consisting of olivine and orthopyroxene with minor plagioclase. Iron-nickel metal makes up about half of its composition. This piece measures 3.5 cm. Specimen supplied by Bruno Fectay and Carine Bidaut, The Earth's Memory, meteorite.fr.

Figure 5.7. NWA 725, found in Morocco, was originally classified as an acapulcoite, but recent oxygen isotope studies suggest that it is a winonaite. This stone, one of 11 found, measures 13 cm. Specimen supplied by Bruno Fectay and Carine Bidaut, The Earth's Memory, meteorite.fr.

Figure 5.8. A slice of NWA 725 shows metal. This meteorite contains relict chondrules which suggests winonaites are actually metachondrites (metamorphosed chondrites). The cube on the left is one centimeter on a side. Courtesy of Eric Twelker, www.meteoritemarket.com.

Asteroidal Achondrites

Basaltic achondrites: Eucrites (EUC)
Fall and find statistics: Falls—23; Finds—193
Well-known examples: Camel Donga; Ibitira; Millbillillie, Pasamonte

Eucrites are the most common of the achondrites and account for about 5% of all known meteorites. About 3% of these are recorded as witnessed falls. Roughly 52% of HED meteorites are eucrites. They may be distinguished from other stony meteorites by their shiny dark brown to black fusion crust which gives them a glassy luster. Eucrites lack chondrules suggesting that these achondrites have a different history than the chondrites. Eucrites are made up of fine-grained fragments composed of igneous rock; that is, rock that formed under magmatic conditions much like terrestrial basalt. Basaltic eucrites have been compared to terrestrial basalt but at first glance, they don't look at all the same. Eucrite interiors are light gray in color compared to the dark gray of terrestrial basalt. The presence of the medium light gray clinopyroxene mineral, *pigeonite*, contributes to the eucrite's lighter interior hue. On Earth, terrestrial basalt is usually dark because it is composed of the iron-rich clinopyroxene mineral, *ferroaugite*, giving the entire rock a dark gray to nearly black color. Eucrites are calcium-rich. They are fine-grained volcanic rocks like those seen in terrestrial lava flows but are chemically different from Earth basalt. In addition to pigeonite, the dominant pyroxene, other minerals include sodium-rich plagioclase but not hydrous minerals due to the absence of liquid water.

Millbillillie can be considered a typical eucrite. Viewed under the petrographic microscope the feldspar laths are particularly striking. Large white laths of plagioclase are intergrown with the pigeonite. Much of the interior is made up of fragments of igneous rock broken by impacts on its asteroid parent body.

Eucrites can be very difficult to distinguish from each other. A unique eucrite called Camel Donga was found in 1984 on the Nullarbor Plain, in Western Australia. Unlike other eucrites, Camel Donga is slightly attracted to a magnet due to the presence of small amounts of iron metal, but it is chemically virtually identical to Millbillillie, which is not attracted to a magnet. One way the collector can make this distinction is to use a strong magnet.

Virtually all eucrites show impact brecciation and melting histories—except one. Ibitira, which fell near Minas Grais, Brazil in 1957, is unbrecciated. Some specimens of Ibitira have an undisturbed texture displaying cumulates formed by the settling of crystals in molten magma. In addition, Ibitira is the only eucrite that has a vesicular texture with gas holes averaging about a millimeter in size. The gas holes cover 5–7% of the rock mass by volume.

Eucrites are shown in Figures 5.9–5.14.

Figure 5.9. Stannern, an historic eucrite which fell in 1808 in the Czech Republic, displays a typical shiny black fusion crust and a light colored interior. It is a monomict breccia. Width 5.5 cm. This specimen was supplied by Bruno Fectay and Carine Bidaut, The Earth's Memory, meteorite.fr.

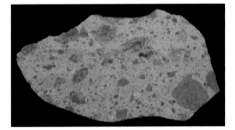

Figure 5.10. Nearly all eucrites are brecciated. NWA 3368 is a typical example of a polymict breccia. A variety of light and dark angular clasts range in size from several millimeters to about 20 mm. They are set in a light matrix typical of eucrites. Width 6.5 cm.

Figure 5.11. This is a fresh block of terrestrial vesicular basalt. Notice the myriad of gas holes permeating the rock. The rock is dark gray to black throughout.

Figure 5.12. All eucrites have light gray interiors. This polymict brecciated slice of Millbillillie has been highly polished to show darker clasts that feature glomerocrysts (clumps of large crystals) with white blades of plagioclase. Slice measures 4.3 cm.

a

b

Figure 5.13. a Thin section of terrestrial olivine basalt seen through a petrographic microscope. The long gray blade-like crystals are plagioclase feldspar. In between the feldspar is olivine and orthopyroxene. **b** Thin section of Millbillillie eucrite shows similar long gray blades of plagioclase feldspar with colorful clinopyroxene demonstrating that eucrites are pieces of a basaltic lava flow.

Figure 5.14. Unlike most basaltic eucrites, Ibitira, which fell in Brazil in 1957, is unbreacciated. It is the only eucrite which displays vesicular texture with gas holes 1 mm or greater in diameter covering the entire specimen. For this reason some researchers feel it may come from a parent body other than Vesta. Perhaps the Dawn mission will resolve this question when the spacecraft arrives at Vesta in 2011. Width of slice is 30 mm.

Basaltic achondrite: Diogenites (DIO)
Fall and find statistics: Falls—11; Finds—156
Well-known examples: Bilanga, Johnstown, Tatahouine

Diogenites are named for the fifth century B.C. Greek philosopher, Diogenes of Apollonia, considered to be the first person to suggest that meteorites actually come from beyond Earth. They are called plutonic since their origin appears to be plutonic rocks deep below the eucrite crust of the asteroid 4 Vesta.

Mineralogically, diogenites are monomineralic, primarily composed of almost pure coarse-grained orthopyroxene (iron-rich hypersthene and bronzite) with small amounts of magnesium-rich olivine and plagioclase feldspar (anorthite). The pyroxene texture is easily seen through a simple hand lens. The large grains probably grew as cumulates deep within the intrusive magma chambers growing within the parent body. Nearly all of the diogenites are monomict breccias. This is best seen looking at a cut slab of the Johnstown, Colorado diogenite, one of the rarest and best preserved. The breccias are composed of large pale green angular fragments ranging in size of between 0.01 and 25 mm or more. Viewed in thin section in cross-polarized light, Johnstown is a wonder to behold.

Johnstown is among the best known and most popular of the diogenites. This is not entirely because of its scientific or commercial value. This meteorite fell on July 6, 1924. It happened that on that very afternoon people were gathering for a funeral service in front of a church two miles west of Johnstown, when suddenly out of the blue came a series of explosions followed by the fall of black stones. One struck near the doors of the church where the service was being held. A half hour after the service had ended the church undertaker picked up a 15-lb black stone. What a send-off!

Diogenites are shown in Figures 5.15–5.18.

Figure 5.15. This is Johnstown, found in Colorado in 1924, one of the best preserved diogenites. Large hypersthene crystals (iron-rich enstatite) are set in a matrix of lighter coarse-grained hypersthene. This slice measures 5.4 cm.

Figure 5.16. This heavily fractured and brecciated Bilanga diogenite fell in Africa in 1999. The sheared and broken character of this hypersthene meteorite is well-illustrated by the broken and offset dark layer. Width of photo 30mm.

Figure 5.17. A thin section of the diogenite Bilanga. Hypersthene of all sizes dominates this portrait of an entire thin section. In the right half, trails of pulverized hypersthene occupy areas between large crystals. Cross-polarized light. Courtesy of John Kashuba.

Figure 5.18. A complete thin section of Tatahouine, a diogenite. Shock-induced cleavage lines and fractures are clearly evident in these large crystals of hypersthene. A brownish mixture of pulverized hypersthene has been injected between some of the large hypersthene crystals. Cross-polarized light. Courtesy of John Kashuba.

Basaltic achondrites: Howardites (HOW)
Fall and find statistics: Falls—20; Finds—156
Well-known examples: Blalystok, Kapoeta, NWA 1929, Pavlovka

These meteorites are named for Edward C. Howard, the early nineteenth century British chemist and meteorite pioneer, whose work on the historic Wold Cottage fall of 1795 helped convince skeptical scientists that rocks do indeed fall from the sky.

Like the other members of the HED association, howardites most likely come from the asteroid 4 Vesta. They are cemented soils composed of clasts and fragments of eucritic and diogenitic material. They are always polymict breccias, often containing black clasts of carbonaceous chondrite material and xenolithic inclusions. (Xenoliths are inclusions of a foreign rock trapped within an igneous rock.) Probably they represent the rocky debris excavated by a large impact that created the enormous crater at Vesta's south pole. This mixture of pulverized eucritic, diogenitic and foreign materials formed a regolith similar to the consolidated soils found on the Moon as well as on many of the asteroids. A regolith is produced by repeated asteroidal impacts resulting in the reworking of the surface by a process known as gardening. In essence, these are the soils of airless worlds.

Like eucrites, howardites have black shiny fusion crusts, the result of a high calcium composition (clinopyroxene). Howardites are about as rare as diogenites.

Howardites are shown in Figures 5.19–5.22.

Figure 5.19. NWA 1929, a howardite found in Morocco in 2003. Howardites are breccias made up mainly of eucritic (basaltic) clasts with a smaller proportion of diogenitic material and melt clasts. This meteorite has been highly shocked and recrystallized. Width 45 mm.

Figure 5.20. Kapoeta, a howardite from Sudan. Interior shows large diogenitic (hypersthene) clasts and white eucritic clasts mixed with fine-grained material of the same composition. Black grains are xenolithic fragments of a carbonaceous chondrite.

Figure 5.21. Kapoeta thin section seen in cross-polarized light reveals a thick howardite area (top and left) wrapping around a wedge-shaped clast (right). There is a dark rim between them with a clast-filled impact melt layer of needle-shaped pyroxene crystals. Width of view 10 mm.

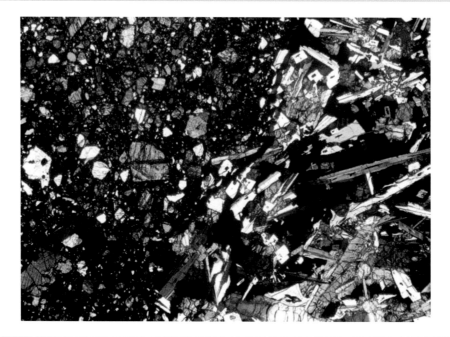

Figure 5.22. Thin section of NWA 2226, a howardite. Two large clasts are juxtaposed. In the left half of the photo is a clast of diogenite with abundant, small broken pieces of hypersthene. In the right half is a piece of a basaltic lava flow (eucritic material) with long laths of white and gray plagioclase and colorful clinopyroxene. Cross-polarized light. Courtesy of John Kashuba.

Asteroidal achondrite: Ureilites (URE)
Type specimen: Novo-Urei, fell in Russia, 1886
Weight of Novo-Urei: 1.9 kg

Fall and find statistics: Falls—5; Finds—209
Well-known examples: El Gouanem, HaH 126, Kenna

These rather strange meteorites fell on a farm near the village of Novo-Urei in Central Russia in 1886. Three stones in all actually touched the ground and were temporarily recovered by villagers. One stone weighing 1.9 kg was recovered on the left bank of the Alatyr River at Karamzinka; another on the right bank at Petrovka was later lost; a final specimen landed in a swamp to the south of the Novo-Urei farm and was also subsequently lost. (The fate of one of the lost specimens was unique—it was eaten by the locals!)

Ureilites are among the rarest if not the strangest of the achondrites after the aubrites and diogenites. Over the past 20 years their numbers have increased from about a dozen to over 200 individuals. Ureilites are totally unique and have little in common with the other achondrites. They are igneous rocks composed of olivine, clinopyroxene (pigeonite), iron-nickel metal, and iron sulfide (troilite). Three major types of ureilites have been recognized by researchers: olivine-pigeonite, olivine-orthopyroxene and polymict ureilites. Most are almost completely lacking in feldspar. The most interesting characteristic is the presence of black opaque carbon-rich material filling the inter-grain spaces. This black material is graphite, the low pressure polymorph of carbon. The high pressure polymorph *lonsdaleite*, the hexagonal form of diamond, is often found in the interstices. Diamonds were discovered in the Novo-Urei meteorite in 1888. Some of the silicate crystals show various stages of impact shock. The presence of high pressure carbon polymorphs strongly suggests that these meteorites have suffered impact shock sufficient to transform graphite into diamonds. Thus, the presence of diamonds forming from graphite as a result of severe shock metamorphism implies that ureilites have had a violent impact history. Under room lighting (reflected light), cut slabs appear dark and opaque and just plain uninteresting. Thin sections of ureilites seen in cross-polarized light, however, show a spectacular color field of olivine and pigeonite in various crystal orientations. These "ugly ducklings" of the achondrites are some of the most beautiful of the asteroidal achondrites.

Ureilites are shown in Figures 5.23–5.26.

Figure 5.23. Like most ureilites, this polymict Hughes 009 specimen, found in Australia in 1991, has a dark interior composed primarily of olivine and pigeonite. It also contains clasts of a variety of rock and mineral fragments including feldspar-bearing clasts. Weathered areas are seen as orange. Width 20 mm.

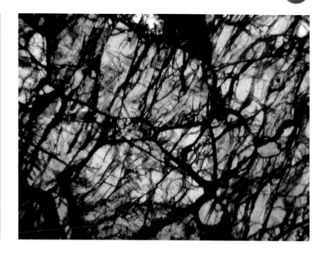

Figure 5.24. In thin section this DaG 319 ureilite shows large equigranular olivine crystals that were fractured during a moderate shock event. Cross-polarized light. Courtesy of John Kashuba.

Figure 5.25. In this thin section of a Dhofar 132 ureilite, colorful olivine crystals have been corroded during a chemical reaction. Fine-grained iron fills the dark areas and corridors where olivine has yielded its iron through a reaction with graphite. Note the 120-degree junction of olivines in the middle of the far right. Cross-polarized light. Courtesy of John Kashuba.

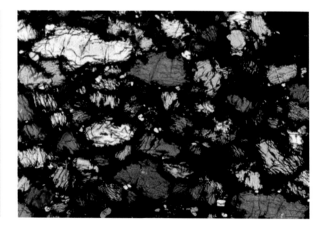

Figure 5.26. In this thin section of a NWA 2624 ureilite, the ragged edges (sawtooth texture) of olivine represent an intermediate stage of corrosion between that seen in Figures 5.24 and 5.25. Cross-polarized light. Courtesy of John Kashuba.

Asteroidal achondrites: Angrites (ANG)
Type specimen: Angra dos Reis, fell in Brazil in 1869
Total weight of Angra dos Reis: 1.5 kg

Fall and find statistics: Falls—1; Finds—9
Well-known examples: D'Orbigny, Sahara 99555

The Angra dos Reis meteorite remained unique among stony meteorites for over a century. Through the years most of the original meteorite disappeared with perhaps no more than 10% distributed among museums. The National Museum in Rio de Janeiro retains the largest specimen (101 g). In three consecutive years (1986, 1987, and 1988) three additional meteorites were discovered in the antarctic ice fields. Two of the three are preserved at the Johnson Space Center in Houston and the final specimen is housed at the National Institute of Polar Research in Tokyo. In May, 1999, the fifth known angrite was found in the Libyan Sahara Desert. This beautiful specimen named Sahara 99555 was a single stone weighing 2.71 kg. It measured about 15 cm in its longest dimension making it the largest angrite in the world. A few slabs have been cut from the original specimen but most remains in private hands. Angrites are ultramafic igneous rocks composed of three calcium-rich primary minerals: plagioclase (anorthite), clinopyroxene, and olivine. Sahara 99555 contains numerous small vesicles a few millimeters across. These cavities may be relic gas bubbles formed before the rock crystallized within the angrite parent body. Curiously, the type specimen Angra dos Reis is unlike the other angrites in that it is monomineralic and probably originated on a different parent body from the other angrites.

In July, 1979, a new angrite was plowed up from a rockless field south of Buenos Aires. The meteorite was kept on the farm as an Indian relic for 20 years. Then, in 1998, the specimen came under suspicion as a possible meteorite. In 2000, a sample was sent to the Natural History Museum in Vienna where it was finally verified as a very rare angrite, now named D'Orbigny. The original weight was 16.55 kg.

D'Orbigny is noted for its coarse-grained interior texture. It shares with Sahara 99555 the presence of round vugs or hollow spheres with thin layers of olivine lining the vugs. Large olivine crystals are scattered throughout the interior and anorthite forms long laths. In thin section, D'Orbigny shows a complex array of mineral crystals in variable states of growth.

Angrites are shown in Figures 5.27–5.29.

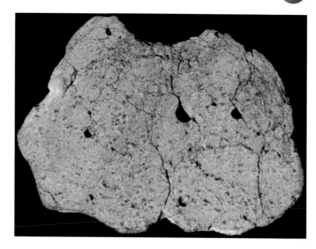

Figure 5.27. Sahara 99555 angrite is among the oldest of meteorites, only slightly younger than CAIs (calcium aluminum inclusions). Their composition is remarkably similar to CAIs, suggesting an origin on a CAI-rich parent body. Note the shadow-filled vesides. Width 15 cm. Specimen provided by Luc Labenne, Labenne Meteorites, www.meteorites.tv.

Figure 5.28. D'Orbigny, the largest angrite ever found, has a very coarse texture with vugs and large olivine crystals. Note the spherical vesicle in the center of the specimen. 0.25g. Courtesy Jeff Kuyken, www.meteorites. com.au.

Figure 5.29. D'Orbigny angrite. Dark and colorful olivine fills the centers and sides of many blades of white and gray plagioclase (anorthite). Augite crystals fill much of the space between plagioclase. The striking and complex look of this beautiful angrite is indicative of its magmatic origin. Courtesy John Kashuba.

Asteroidal achondrites: Aubrites (AUB)
Type specimen: Aubres, fell in France in 1836
Total weight of Aubres: 800 g

Fall and find statistics: Falls—9; Finds—44
Well-known examples: Cumberland Falls, Norton County, Peña Blanca Spring

Aubrites are the only stony meteorites that have a pale beige-colored fusion crust contrasting with a white interior. They lack iron which makes their fusion crusts turn a light tan color. This is a distinguishing characteristic of all aubrites. The meteorites are almost pure magnesium silicate. Until 1948, the aubrites were some of the rarest of the stony meteorites. Then, on February 18, 1948, a fall of over 100 stones occurred over Norton County, Kansas and Furnas county, Nebraska. One weighed over 1 ton and the second most massive weighed in at 131.5 lb. The 1 ton stone remains on permanent display in the meteorite museum at the University of New Mexico in Albuquerque.

Another witnessed fall occurred in 1946 when a meteorite plunged into a small murky pond called Peña Blanca Spring near Marathon, Texas. Over 70 kg were recovered after repeated diving into the pond to retrieve the specimens. Remarkably, a third fall of an aubrite had occurred some years earlier (April 9, 1919) in which 14 kg of aubrite material were recovered near Cumberland Falls, Kentucky.

Aubrites are primarily iron-free enstatite achondrites with minor amounts of troilite (FeS), FeNi metal, plagioclase (calcium-poor), olivine, and clinopyroxene (diopside). They are closely related to the E chondrites, being highly reduced and sharing oxygen isotopic compositions. All but the Shallowater, Texas aubrite found in 1938 are brecciated. Shallowater has an igneous texture. All Aubrites must be handled with some care since they are easily crushed.

Aubrites are shown in Figures 5.30–5.33.

Figure 5.30. This 1-ton aubrite found in Norton County is on permanent display at the University of New Mexico in Albuquerque. Courtesy of Al Mitterling.

Figure 5.31. A piece of the Norton County aubrite. These very rare achondrites are composed of nearly pure enstatite, a magnesium pyroxene. They also contain small amounts of many exotic minerals not found on Earth including the rare sulfides oldhamite and niningerite. Measures 4 cm on the long diagonal.

Figure 5.32. Peña Blanca Spring, an aubrite from Texas. Though subtle, note the various sizes of enstatite fragments and the small brown and orange patches of weathered sulfides. Measures 3 cm on the long diagonal. Courtesy of Anne Black, www.impactika.com.

Figure 5.33. The Cumberland Falls aubrite consists of two kinds of broken angular fragments. One kind is light-colored enstatite, broken and pulverized. The other kind is very dark-colored LL chondritic inclusions from an asteroid which impacted the aubrite parent body. This specimen is in the U. S. National Meteorite Collection in the Smithsonian Institution. Courtesy of Al Mitterling.

Asteroidal achondrites: Brachinites (BRA)
Type specimen: Brachina, found in South Australia in 1976
Total weight of Brachina: 203 g

Fall and find statistics: Falls—0; Finds—12
Well-known examples: Eagles Nest, NWA 595, NWA 3151, Reid 013

These unusual meteorites were first classified simply as anomalous achondrites since they did not fit into any known group. Their mineralogy is quite simple; they are primarily composed of granular olivine (having all mineral grains approximately the same size). When first studied, most meteoriticists thought Brachina was a Chassigny, an exceedingly rare meteorite from Mars. This was understandable for both brachinites and Chassigny were composed of almost pure olivine. But very quickly studies showed that the two were not related. Specifically, like most meteorites, Brachina was as old as the Solar System (4.56 billion years). Chassigny turned out to be a much younger rock (1.3 billion years) made of the same olivine. Brachinites are ultramafic mantle rocks which on Earth are called dunites. About 90% of Brachina is olivine with minor constituents including clinopyroxene (diopside). The meteorite has a total iron content of about 20% with most of the iron locked in the iron-bearing minerals.

Brachinites are shown in Figures 5.34–5.37.

Figure 5.34. Eagle's Nest, an oriented brachinite from Australia, was found by a prospector near an eagle's nest in 1960. This complete slice of the egg-shaped stone displays medium-grained, equigranular crystals of olivine and lesser amounts of orthopyroxene and some sodic plagioclase. Weathering has produced the orange-brown color. Width 6.5 cm. Courtesy of Ken Regelman, Astronomical Research Network, Martin Horejsi, and David Weir, www.meteoritestudies.com.

Figure 5.35. NWA 3151, a brachinite purchased in Morocco in 2005, has equigranular crystals from 0.7 to 1.6 mm in diameter. Courtesy of Greg Hupé, The Hupe Collection, www.LunarRock.com.

Figure 5.36. A thin section of NWA 3151, a brachinite. These millimeter-sized olivine crystals sank and grew together at the bottom of a magma chamber. Cross-polarized light. Courtesy of John Kashuba.

Figure 5.37. NWA 595, a highly weathered brachinite found in 2000. This photo of a complete thin section shows abundant 0.5–1 mm sized olivine crystals with a polygonal-equigranular texture. Minor amounts of orthopyroxene, augite and chromite are present. The layered appearance is probably due to accumulations of olivine crystals that sank to the bottom of a magma chamber. Cross-polarized light.

CHAPTER SIX

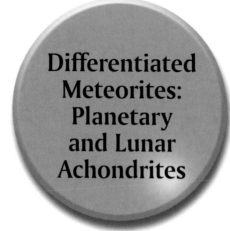

You may recall that virtually all chondrites have similar textures, chemistry and ages of 4.56 billion years. When we examine the achondrites, we find more variety. In the first few million years of Solar System history, the achondrites were either fully melted and differentiated into mineral zones within each parent body or they showed partial melting in an interrupted attempt to differentiate. The heating probably occurred within the first few million years after the chondritic parent bodies accreted. So we see that the ages of the chondrites and these achondrites are approximately the same. But there are also some groups of achondrites which are very different, and these strange meteorites turn out to come not from the asteroid belt but from the planet Mars and the Moon.

Martian meteorites are grouped into three types known by the acronym SNC (pronounced "snick"). A fourth type, represented by one meteorite from Allan Hills in Antarctica was discovered in 1984 and given the name ALH 84001. More will be said of this extraordinary meteorite later. SNC refers to the first letter in each of the names of the three type specimens: "S" is for *shergottite*, named after a meteorite that fell in 1865 in Bihar, India, near the town of Shergotty; "N" is for *nakhlite* after the type specimen that fell in Nakhla, Egypt, in 1911; and "C" is for *chassignite*, named for a meteorite that fell in Chassigny, France in 1815. Most of the SNC meteorites are thought to be cumulate igneous rocks that formed on the floor of a magma chamber.

Totally, there are at least 34 known martian meteorites, and the numbers increase yearly. (This excludes meteorites which are paired, considered to be from the same fall but having different names or numbers.) Different types of shergottites are recognized: currently there are 10 basaltic shergottites; 8 olivine-phyric (porphyritic) shergottites; and 6 lherzolitic peridotite shergottites (lherzolitic means ultramafic ferromagnesian plutonic rocks). In addition, there are 7 clinopyroxenites (nakhlites); 1 orthopyroxenite (ALH 84001); and 2 dunites (chassignites). Of the 34 martian meteorites known only four were seen to fall. These are Shergotty, Zagami, Nakhla, and Chassigny. An excellent web site on martian meteorites is provided by the Jet Propulsion Laboratory in Pasadena, California at: www2.jpl.nasa.gov/snc.

Martian SNC group

SNC—Shergottites
Type specimen: Shergotty, fell in India in 1865
Total weight of Shergotty: 5 kg

Fall and find statistics: Falls—2; Finds—22 (excluding paired specimens)
Well-known examples: LA 001 and LA 002 (paired), Zagami

In October 1999 two meteorites found in the Mojave Desert of southern California some 20 years earlier were taken to the University of California at Los Angeles where researchers quickly identified them as meteorites. But more significantly, the researchers concluded that they were paired martian meteorites, specifically, shergottites. Several indicators brought them to that conclusion. First, the gas isotopic composition was unique to the martian atmosphere. Second, the hydrogen to heavy hydrogen ratio was high. This is because Mars, only 11% the mass of Earth, cannot gravitationally hold onto the lighter hydrogen which escapes into space leaving the heavier deuterium behind. Their findings were all consistent with measurements made by the Viking Mars lander in 1976. Most significantly, the meteorites had crystallization ages of only 1.3 billion years, far too young to be asteroidal. And their age matched the age of several other shergottites in laboratories on Earth.

All shergottites show substantial signs of shock metamorphism leading to glass formation. Their augite and feldspar exhibit undulatory to mosaic extinction. These shock features can be seen in the crystals making up the shergottite called Zagami, which fell in Nigeria in 1962. Shergottites are the most common of the SNC group. Most have basaltic compositions with pigeonite, augite, and maskelynite as major minerals. Maskelynite, a type of glass that has the composition of plagioclase feldspar, was created by the vitrification of the plagioclase by shock melting during impacts. The maskelynite phase comprises about 23 vol.%. It is isotropic, meaning that it remains in extinction under cross-polarized light in all optical rotation angles when viewed in a petrographic microscope. (More on this in Chap. 11.)

Shergottites are shown in Figures 6.1–6.4.

Figure 6.1. A slice of the Zagami shergottite which fell in Nigeria in 1962. This strongly shocked piece of basaltic lava was ejected from Mars during an impact event about 3 million years ago. It consists of 75% fine-grained pigeonite and augite, 18% plagioclase glass (maskelynite) and several minor minerals. Except for being shocked, it is similar to terrestrial basalts. Width 33 mm.

Figure 6.2. Sayh al Uhaymir (SaU) 130, a shergottite found in Oman in 2004. It is a porphyritic basalt with large brown crystals of olivine surrounded by a light-colored, fine-grained groundmass of pigeonite and maskelynite. Widest dimension 38 mm. Courtesy of Jim Strope, www.catchafallingstar.com.

Figure 6.3. A thin section of SaU 008, a shergottite paired with SaU 130, seen in Figure 6.2. Elongated crystals of white to orange twinned pigeonite and small dark areas of maskelynite are surrounded by colorful phenocrysts of olivine. Cross-polarized light. Longest dimension 16 mm. Courtesy John Kashuba.

Figure 6.4. NWA 1950, a lherzolitic peridotite found in 2001 in Morocco, an ultramafic rock of plutonic origin. It consists mainly of olivine (about 55 vol.%), plus Ca-poor and Ca-rich pyroxene (about 35%). Width 34 mm. Courtesy of Jim Strope, www.catchafallingstar.com.

SNC—Nakhlites
Type specimen: Nakhla, fell in Egypt in 1911
Total weight of Nakhla: 10 kg

Fall and find statistics: Falls—1; Finds—8
Well-known examples: Governador Valadares, Lafayette

Until recently, only three nakhlites were known. These include specimens from Nakhla, Egypt, Lafayette, Indiana, and Governador Valadares, Brazil. In December 2000, a 104 g specimen was discovered in Morocco, called Northwest Africa (NWA) 817, bringing the total nakhlites to four. In the same year, an enormous nakhlite find was made in Antarctica by Japanese scientists totaling 13,731 g. The nakhlite NWA 998 was found in 2001 in Morocco. It is extremely important because of its similarities to the ALH 84001 stone, the enigmatic meteorite which may indicate the early presence of life on Mars. It contains water bearing minerals and orthopyroxene crystals.

Augite is the primary pyroxene cumulate mineral in nakhlites. It amounts to about 80 wt.% and gives the interior of the stones a greenish cast. In thin section the augite grains are long prisms and, as in shergottites, often show simple twinning. All nakhlites contain the alteration product known as *iddingsite*, which is often seen running through veins in olivine indicating the presence of water. Unlike the shergottites and the chassignites, nakhlites show only minor signs of shock.

Nakhlites are shown in Figures 6.5–6.8.

Figure 6.5. NWA 998, the front and back of a nakhlite with a beautiful shiny black fusion crust (left image) and an interior dominated by the mineral augite (right image). It contains water-bearing minerals, reinforcing the idea that Mars was once wet and may have been able to support life. Height 34 mm. Courtesy of Jim Strope, www.catchafallingstar.com.

Figure 6.6. An entire thin section of a NWA 998 nakhlite made up of equant (generally having equal diameters in all directions) augite crystals. Cross-polarized light. Courtesy of John Kashuba.

Figure 6.7. A detail of a thin section of NWA 998. It shows a feast of augite crystals. These clinopyroxene crystals probably sank and accumlated at the bottom of a Martian magma chamber. Note twinning where crystal halves are different colors. Fine, light, comb-like lines are exsolution lamellae in the dark gray and blue central elongated crystal. Courtesy of John Kashuba.

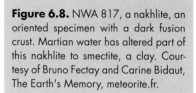

Figure 6.8. NWA 817, a nakhlite, an oriented specimen with a dark fusion crust. Martian water has altered part of this nakhlite to smectite, a clay. Courtesy of Bruno Fectay and Carine Bidaut, The Earth's Memory, meteorite.fr.

SNC—Chassignites
Type specimen: Chassigny, fell in France in 1815
Total weight of Chassigny: 4 kg

Fall and find statistics: Falls—1; Finds—1
Well-known example: Chassigny

The only chassignite seen to fall, and for nearly two centuries the only chassignite known, fell near the village of Chassigny in Haute-Marne province, France. Mineralogically, it is composed of about 90% iron-rich olivine. When first found, it was mistaken for a brachinite which is also composed of nearly 90% iron-rich olivine. It is similar to terrestrial dunites. Dunite is a variety of peridotite being composed almost entirely of olivine along with minor amounts of pyroxene, plagioclase and chromite. Much of the feldspar is highly shocked (S5) and occurs as diaplectic glass.

Finally, a second chassignite (NWA 2737) was found in Morocco in August, 2000. It weighed about 600 g. It too is a cumulate rock composed of dunite, much like the original type specimen. Meteorite collectors Bruno Fectay and Carine Bidaut brought several mysterious black rocks from Morocco that resisted a quick identification. Not recognizing them as meteorites, they placed the rocks in a meteorwrong pile for possible future study. During the summer of 2004 a small piece of the stones that would become NWA 2737 was given to scientists for analysis. Before it was given an official name, they called it Diderot, in homage to the eighteenth century French encyclopedist. The specimens (nine in all) resemble a terrestrial cumulate dunite, much like the original Chassigny. They contain about 90% iron-rich olivine, 5% clinopyroxene and 1.7% plagioclase plus minor mineral components. NWA 2737 is highly shocked (S5), which turned the normally yellow olivine a blue-black color. This deformation in olivine has not been observed before. Its crystallization age is 1.36 by, much like the other Mars meteorites and much younger than any of the chondrites.

Chassignites are shown in Figures 6.9–6.11.

Figure 6.9. Bruno Fectay and Carine Bidaut hold samples of their new chassignite find found in the Moroccan desert. This NWA 2737 was originally called Diderot. Courtesy of Bruno Fectay and Carine Bidaut, The Earth's Memory, meteorite.fr.

Figure 6.10. NWA 2737, only the second chassignite ever found. It was recognized in 2004. Specimen supplied by Bruno Fectay and Carine Bidaut, The Earth's Memory, meteorite.fr.

Figure 6.11. Piece of the chassignite NWA 2737 reveals darkened olivine caused by intense shock (S5). This piece measures 34 mm. Courtesy of Bruno Fectay and Carine Bidaut, The Earth's Memory, meteorite.fr.

ALH 84001

Non-SNC orthopyroxenite—ALH 84001
Found in Antarctica in 1984 at Allan Hills
Total weight of ALH84001: 1.9 kg

Both martian and lunar meteorites are associated with extraordinary stories of success and failure. One involves the most famous meteorite in the world, the Antarctic meteorite ALH 84001. When Roberta Score of the U.S. Antarctic Meteorite Recovery team collected this meteorite from the ice fields on December 27, 1984, several researchers noted its olive-green color that reminded them of the orthopyroxene mineral, *hypersthene*. Specifically, from its color alone it had the appearance of a relatively common diogenite from the HED group. And that is the way it was classified. The meteorite remained in storage at the Johnson Space Center in Houston for nearly 10 years, mislabeled and all but forgotten. Then, in 1993, David Mittlefehldt, a researcher at the Johnson Space Center, looked closer and found tiny grains of trace minerals normally not seen in diogenites. Further study showed that the meteorite was related to the SNC group but with a composition representing a new SNC type. ALH 84001 had an extraordinary history. It had a coarse crystalline texture showing that it had crystallized from a melt deep within the martian crust, in a location now believed to be in the Eos Chasma, a branch of the enormous Valles Marineris canyon. Shocks from nearby impacts produced fractures throughout the rock. Then water, charged with CO^2, seeped through the veins depositing tiny beads of carbonate minerals. An asteroid impact ejected ALH 84001 into the inner solar system. For 16 million years the meteorite crossed Earth's orbital path finally plunging into Earth's atmosphere and landing in Antarctica where it waited another 13,000 years for someone to retrieve it. That "someone" was a team of nine researchers led by David S. McKay and Everett K. Gibson, Jr.

McKay's team studied strange structures hiding within the green crystals. Carbonate deposits were found in unlikely association with magnetite and FeS. There were organic compounds typical of decaying organic material on Earth and strange fossil-like structures closely resembling fossil bacteria found on Earth in three billion year old terrestrial rock. After an intense two year study, the group announced to the world on August 7, 1996, that they had found evidence for microbial life in ALH 84001. The carbonate deposits became a source of great interest. These deposits are yellow-orange globules averaging about 50 μm across and rimmed with alternating black and white magnesium and iron-rich layers. Between the layers they found tiny magnetite and iron sulfide grains in close proximity. The carbonate deposits also contained organic compounds called polycyclic aromatic hydrocarbons or PAHs. PAHs are large organic molecules often associated with life processes on Earth. They form on decaying organic matter. At the same time, a more provocative statement made in McKay's paper was the announcement that bacteria-like structures had been found in the carbonate globules. These appeared to be identical to terrestrial bacilli except that they were only about a tenth the size of the smallest bacteria known on Earth.

So the debate continues, probably not to be solved until manned or unmanned spacecraft reach Mars, search for the evidence, and return to Earth a precious cargo of martian rocks for further study.

ALH 84001 is shown is Figures 6.12–6.14.

Figure 6.19. NWA 482. This is a shocked lunar highland rock, an anorthositic impact melt polymict breccia. Note the dark-colored glassy and vesicular melt veins and melt pockets. The light interior is due to scattered white clasts of anorthosite set in a fine-grained matrix of anorthitic plagioclase. Width 8 cm. Courtesy of Mike Farmer, www.meteoritehunter.com.

Figure 6.20. NWA 4472, a lunar breccia. This meteorite is made of various types and sizes of light to dark basalt fragments. Many of the smaller fragments are individual broken crystals of pyroxene, olivine, plagioclase, metal (both kamacite and taenite), troilite and many others. Block is a 1-cm cube. Courtesy of Greg Hupé, the Hupé Collection, www.LunarRock.com.

Figure 6.21. A piece of NWA 032, an olivine-pyroxene lunar basalt found in Morocco in 2000. Yellow-white fragments are fractured olivine. Matrix consists of small crystals of pyroxene and chromite. Courtesy of Bruno Fectay and Carine Bidaut, The Earth's Memory, meteorite.fr.

Figure 6.22. NWA 773 was found in 2000 in Western Sahara. A meteorite from the lunar highlands classified as a lunar regolith gabbro. It is a piece of the lunar regolith containing two large clasts of olivine gabbro and darker gray regolith breccia. Specimen courtesy of Marvin Killgore.

Figure 6.23. NWA 2727 (paired with NWA 773). A regolith breccia with two clasts of gray basalt and abundant small fragments of gabbro. Width 20 mm. Courtesy of John Kashuba.

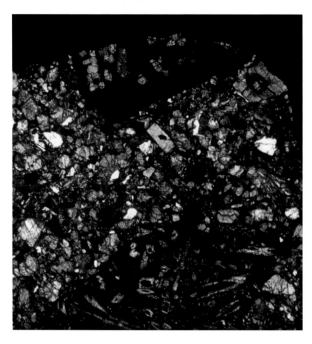

Figure 6.24. A low magnification thin section of the right half of NWA 2727 in Figure 6.23. The two clasts of basalt are black areas with colorful scattered crystals of pyroxene. Between the basalt clasts are abundant angular fragments rich in crystals of pyroxene and anorthite. Cross-polarized light. Courtesy of John Kashuba.

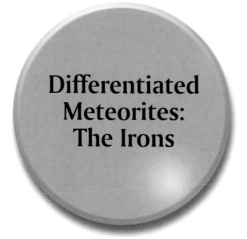

CHAPTER SEVEN

Differentiated Meteorites: The Irons

In Chap. 4, we had our first look at the chondrites. We saw that they are among the most primitive bodies in the Solar System and many contain spherical silicate bodies called chondrules. They are relatively simple bodies made up of only a few minerals that were first to condense out of the solar nebula. Along with chondrules, the matrix enclosing these spherical bodies housed tiny grains of metal, primarily iron alloyed with nickel. These were precursor metal grains that formed along with other refractory (high temperature) grains, melting and recrystallizing within the solar nebula. In time, these "chondritic" bodies accreted, growing larger and more massive by the assimilation of smaller bodies that collided with the larger parent. After reaching anywhere from 6 to 200 km in diameter, the chondritic parent body melted and differentiated, resulting in a layered planetary body complete with crust, mantle and core. The asteroid 4 Vesta is an example of such a differentiated body.

Early in the evolution of these asteroid parent bodies, the crusts and mantles began to be striped away by further impacts, exposing their metallic cores. Continuing impacts over millions of years must have completely denuded many of these asteroids to the core. The M-type asteroid (M for metal) 16 Psyche is a well-known asteroid that shows a flat featureless spectrum, reflecting about 10% of the light striking it. At the present time we can only observe these metallic bodies from great distances for the core of a differentiated asteroid is inaccessible unless it is completely shattered by another impacting body. For centuries, we have been blessed with bits and pieces of iron falling from the sky. If achondrites come from the crusts and mantles of other worlds, then iron meteorites must represent the deep foundation of these worlds. These are the irons—the cores of differentiated planets.

Alloys of Iron-nickel Meteorites

The two most important iron-nickel alloys found in iron meteorites are low-nickel kamacite or *alpha-iron* and high-nickel taenite or *gamma-iron*. These two minerals belong to the isometric crystal system and show hexahedral (cubic) crystals. In a metallic melt, both kamacite and taenite begin to form as the liquid metal cools below 1370 °C. Which alloys form depends upon the nickel content of the melt, the temperature at the time of crystallization, and the rate of cooling. When the temperature drops below 900 °C both the taenite and kamacite jointly divide into three fields. Figure 7.1 shows an iron-nickel stability phase diagram that predicts the three stability fields of kamacite, taenite, and kamacite + taenite for various temperatures and nickel compositions. Above 900 °C only the taenite phase of the iron nickel alloy is stable. But as the temperature continues to drop, the stability shifts to the taenite + kamacite field or the kamacite field alone, depending upon the weight percent nickel of the alloy. For a melt containing more than about 30 wt.% nickel, only the taenite structure exists. Diffusion of nickel stops before the kamacite structure can form. On the kamacite or low nickel end with a melt of 5% nickel, all the taenite changes to kamacite before diffusion stops. Between 6% and 13% nickel the mineral composition contains both alloys, i.e., there is an intergrowth of both alloys.

Why should we be concerned about the stability of these alloys? Simply because this difference in composition leads to visible structural differences by which they can be classified. Through the years meteoriticists have devised two methods of classifying iron meteorites. The older method is based upon the characteristic crystalline patterns that appear when an iron-nickel meteorite is polished and etched. However, this applies only to the octahedrites, the most common type of iron meteorite. They have less than 6 wt.% nickel. When etched, the meteorite reveals a characteristic pattern of lamellae (plates) of kamacite intergrown with nickel-rich phases. The width of the kamacite plates allows octahedrites to be classified into six structural subgroups.

Since the mid 1950s a chemical method of classifying iron meteorites involving measuring the presence of trace elements such as gallium, germanium, and iridium in microscopic samples has been developed. Concentrations of these trace elements are plotted against the overall nickel content on a logarithmic scale. Originally there were four groups recognized but now there are

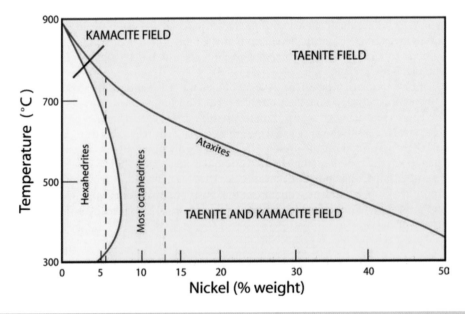

Figure 7.1. Iron-nickel stability phase diagram shows the three stability fields of kamacite, taenite, and kamacite + taenite for various temperatures and nickel compositions.

Figure 7.2. Calico Rock, a IIAB HEX, was found in Arkansas in 1938 and weighed 7.3 kg. It was almost perfectly rectangular in shape (a single cubic crystal). This specimen is covered completely by Neumann lines in three sets of parallel lines. Width 47 mm.

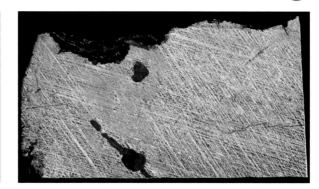

Figure 7.3. Richland (formerly called Fredericksburg), Texas, a IIAB HEX, found in 1951. Fredericksburg was considered to be a separate fall but has been shown to be the same as Richland, though they were found 300 km apart, no doubt moved by human activity. This slice measures 120 × 63 mm.

Figure 7.4. Boguslavka IIAB HEX. This single hexahedrite crystal of kamacite shows three sets of Neumann lines. Note the recrystallized outer layer due to heat alteration during passage through the atmosphere. Longest dimension 24 mm. Specimen courtesy of Tim Heitz, Midwest Meteorites.

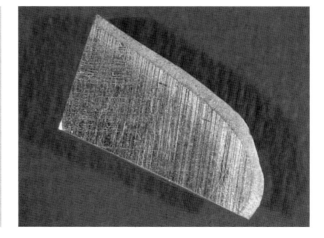

Group: Octahedrite (O)
Subgroups: Coarsest octahedrite (Ogg): Lake Murray, Sikhote-Alin
Coarse octahedrite (Og): Canyon Diablo, Odessa
Medium octahedrite (Om): Henbury, Cape York
Fine octahedrite (Of): Gibeon
Finest octahedrite (Off): Glen Rose
Plessitic octahedrites (Opl): Taza (NWA 859)
Predominate structure: Widmanstätten (Thomson) figures; some Neumann lines on kamacite bands

The intergrowth of kamacite and taenite produces some of the most remarkable and beautiful structures found in any class of meteorites—the Widmanstätten structure. To see these octahedral figures appear is something almost magical. The octahedral figures were first discovered by William Thomson in Naples in 1804 and independently four years later by Count Alois von Widmanstätten in Vienna. Thomson was trying to keep the Pallas stony-iron from rusting, using nitric acid as a preservative. He failed to keep the rust from corroding the meteorite but at the same time the acid etched the specimen, bringing out the octahedral pattern. Count von Widmanstätten was investigating the properties of iron meteorites in 1808 when he accidently produced the figures as he was heating a thin octahedrite slab using a Bunsen burner. During the heating process the kamacite and taenite oxidized the meteorite at different rates, thus bringing out the pattern.

Kamacite grows at specific sites on the taenite cube as it slowly changes from a face-centered cubic structure to a body-centered structure. In Figure 7.5 on the left we see taenite in its hexahedral form with all three axes equal. Kamacite plates begin to grow on the taenite crystal by truncating the corners of the cube at 45 degree angles. As growth of the kamacite plates continues, the truncated corners finally meet at all three axes on opposing sides of the cube. The new form on the right is an eight-sided dipyramidal figure made up of eight equilateral triangles. This new crystal shape is still based upon the cube. The eight-sided figure is an octahedron and gives its name to the most common irons, the octahedrites.

Most meteoriticists still use the structural classification scheme based upon the texture of the Widmanstätten structure or the width of the kamacite plates which, in turn depends upon the bulk nickel content. The octahedrite classification is composed of *six subgroups*: Ogg—coars-

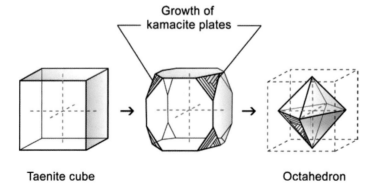

Growth of kamacite plates

Taenite cube   Octahedron

Figure 7.5. The cubic crystal form of iron-nickel changes from a hexahedron (taenite) to an octahedron (kamacite) as the proportion of nickel increases. New faces grow at the corners of the hexahedron and replace the faces of the cube.

est (Figure 7.6–7.7); Og—coarse (Figure 7.8–7.12); Om—medium (Figure 7.13–7.15); Of—fine (Figure 7.16–7.19); Opl—Plessitic (Figure 7.20–7.21).

Octahedrites are the most plentiful of the iron meteorites extending from the coarsest (Ogg) to the finest (Off). The coarsest bands have the least bulk nickel but there is substantial overlap of nickel content for each subgroup. Beyond the finest subgroup are the plessitic octahedrites. Plessite is a fine-grained mixture of kamacite and taenite. Here, the Widmanstätten structures give way to kamacite spindles in a plessitic granular matrix when the nickel content reaches about 13%. These in turn grade smoothly into ataxites, a structureless mass of high-nickel taenite with a nickel content of over 16%.

Octahedrites are shown in Figures 7.6–7.21.

Figure 7.6. Lake Murray, a IIAB Ogg iron with a bandwidth of 10 mm. This meteorite was discovered in 110-million-year-old Lower Cretaceous beds in Oklahoma in 1933. It has the oldest terrestrial age of any meteorite known. Note swathing kamacite which surrounds a troilite nodule on lower left. Individual plates have Neumann lines. Width 23 cm.

Figure 7.7. A slab of Arispe, a IC Og iron, found in Sonora, Mexico in 1896. Bandwidth is 2.9 mm. Some bands of kamacite exhibit Neumann lines. Longest dimension 5.2 cm.

Figure 7.8. Canyon Diablo, a IAB Og iron meteorite found near Barringer Crater, northern Arizona. Note well-developed regmaglypts with sharp-edged boundaries on this 7 pound specimen. The small orange-red patches are due to terrestrial weathering.

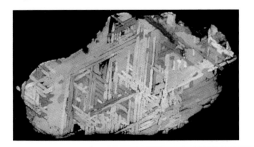

Figure 7.9. A slice of the Toluca IAB Og iron found in 1776 in Xiquipilco, Mexico. Bandwidth is 1.4mm. The specimen was cut parallel to an octahedral crystal face, forming 60° and 120° angles in the Widmanstätten figures. Note bronze-colored, elongated troilite inclusion in upper right. Longest dimension 10cm.

Figure 7.10. This Campo del Cielo iron, a IAB Og, was found in 1576 in Argentina. Bandwidth is 3.0mm. The darkest patches are silicate inclusions commonly found in this meteorite. Note Neumann lines in the darker plates. Length 11cm.

Figure 7.11. A part slice of the famous Odessa IAB Og iron from Ector County, Texas, discovered in 1922. Schreibersite surrounds dark graphite inclusions of various sizes. This piece is 8cm × 7.3cm.

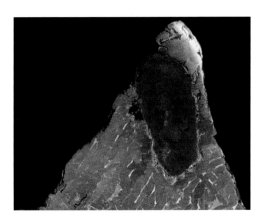

Figure 7.12. A detail showing the graphite nodule in the Odessa IAB Og in Figure 7.11. The nodule is surrounded by schreibersite. Rhabdite is the old name for schreibersite. Measures 2.4 × 1.5cm.

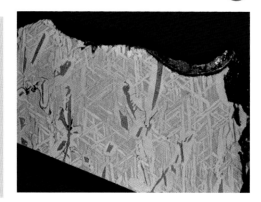

Figure 7.13. Tambo Quemado, a IIIB Om iron found in Peru in 1949. It has 8 wt % nickel and kamacite bandwidths of 0.7 mm. This meteorite was probably reheated to about 1000 °C at some point in its past, which resulted in melting of the schreibersite inclusions (medium bronze color). Field of view is 6.5 cm long.

Figure 7.14. A slice of the Cape York IIIAB Om iron found in Greenland in 1818. This piece is from the mass known as Agpalilik, which weighed about 15 tons. Bandwidth 1.2 mm. Note the rectangular fields of plessite (dark areas) among the kamacite bands. Width 5.8 cm.

Figure 7.15. Zacatecas, a IIIAB Om iron, found in Mexico in 1969. It shows the effects of shock heating and recrystallization in the lower right corner. It has 5.9% nickel. Bandwidth 0.7 mm. Measures 7 × 7 cm.

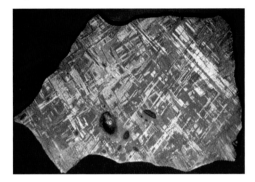

Figure 7.16. Gibeon, a IVA Of iron found in 1838 in Namibia. Bandwidth varies from 0.3 to 0.5. Ni content is 7.9%. The Gibeon strewn field covers an area 70 × 230 mi, running SE to NW. More than 25 tons have been found. No craters have been located. One of the most prized of the octahedrites. Width 17.5 cm.

Figure 7.17. Detail of the Gibeon slab in Figure 7.16. The large dark inclusion is an excavated (hollow) troilite nodule which shows evidence of partial melting. The nodule is 1.4-cm long. Width of image 6.5 cm.

Figure 7.18. A slice of the beautiful Rica Aventura IVA Of iron from Chile, found in 1910. Bandwidth 0.27 mm, nickel 8.9%. Weighs 215 g. Photograph by Geoffrey Notkin/Aerolite. org, © The Michael Farmer Collection/ www.meteoritehunter.com

Figure 7.19. Muonionalusta, a IVA Of iron from Sweden, found in 1963 at a building site. Bandwidth 0.3 mm, nickel 8.4%. This meteorite displays subtle evidence of shearing by shock displacement in the kamacite plates. Longest dimension 14.6 cm.

Figure 7.20. NWA 859, also called Taza, an Opl iron meteorite found in Morocco in 2000. A pattern of kamacite spindles (laths) is seen against the plessitic matrix. Longest dimension 37 mm.

Figure 7.21. A detail of NWA 859 (Taza) showing microscopic spindle-shaped laths of kamacite surrounded by taenite. The fine-grained metal mixture called plessite, which consists of an intergrowth of kamacite and taenite, fills the triangular spaces between the laths.

Group: Ataxite (D)
Predominate structure: not resolvable to the naked eye but resolved into microscopic Widmanstätten patterns, some silicate inclusions
High nickel: >16%
Kamacite: only minute quantities
Band width: none
Well-known examples: Hoba, Santa Catharina, Tucson Ring

Ataxites were once divided into two groups: nickel-rich and nickel-poor. This nomenclature is no longer used today. The name ataxite is from the Greek meaning "without structure," a good name since early researchers were not able to detect any internal structure. The designation D is from the German *Dichte Eisen*, meaning dense iron. Though Widmanstätten patterns are not easily seen with the naked eye, a fine microscopic structure is visible which shows tiny taenite crystals coated with laths of kamacite and set in a fine plessitic matrix. These meteorites consist almost entirely of taenite.

Ataxites are shown in Figures 7.22–7.24.

Figure 7.22. A nugget of the ataxite Chinga, a IVB D iron found in Russia in 1912. Chinga shows evidence of shock both in space and as a result of its energetic impact. Pieces of Chinga are usually flattened and angular as a result. Dimensions 45 × 50 mm.

Figure 7.23. A polished slab of Chinga, a IVB D iron, which has a high nickel content of about 18% and has resisted rusting even though it fell into a riverbed more than 2000 years ago. No structure can be seen in this meteorite even when etched. Ataxites are generally polished to a mirror surface and are not etched. Longest dimension 7 cm.

CHAPTER EIGHT

Differentiated Meteorites: Stony-Irons

You recall that meteoriticists once divided meteorites into three primary classes: stones, irons, and stony-irons. As we saw in Chap. 4, the vast majority of meteorites were classified as primitive chondrites. This simple division, though convenient, quickly proved to be inadequate. Because of thermal metamorphism in the early Solar System, internal heating on some of the more massive parent asteroids produced a broad range of meteorite types from the original precursor chondrites. Those meteorite types with low degrees of melting, such as the acapulcoites and lodranites, produced an imperfect separation of metal and silicate materials while higher grades of melting produced shadowy relics of their chondritic past history.

Stony-irons, which are composed of half iron and half silicates, are divided into two main groups: pallasites and mesosiderites. The pallasites formed at the boundaries between the metal cores and silicate mantles of melted and differentiated asteroid parent bodies. Mesosiderites are also stony-irons that have roughly equal portions of metal and silicates. Beyond that, the mesosiderites seem to have little in common with the pallasites. They were formed by impact melting, the result of violent collisions in the early Solar System and represent a mixture of a wide sample of parent bodies.

Pallasites

> Group: Main Group Pallasite (MGP)
> Type specimen—Main Group: Krasnojarsk, Siberia (Pallas iron), found in 1749
> Weight of Krasnojarsk: 700 kg
>
> Fall and find statistics: Falls—4; Finds—70
> Well-known examples: Brahin, Brenham, Esquel, Glorieta Mountain, Imilac, Marjalahti
>
> Grouplet: Eagle Station (PES)
> Type specimen: Eagle Station, found in 1880 in Kentucky
> Weight of Eagle Station: 36 kg
>
> Fall and find statistics: Falls—0; Finds—3
>
> Grouplet: Pyroxene Pallasites (PXP)
>
> Fall and find statistics: Falls—0; Finds—4

Pallasites are among the most common stony-irons with about 60 known worldwide. The name "pallasite" comes from a 1,600 lb meteorite found in 1749 near Krasnojarsk, Siberia. The German naturalist and explorer, Peter Simon Pallas, collected samples of this unusual stone, an iron containing large olivine crystals, and described it in his journals in 1772. Pallas had been invited by the Russian Monarch, Catherine the Great, to explore the vast uncharted regions of the Siberian taiga (Mount Emir), especially around Krasnojarsk. He didn't realize at the time that he had found a meteorite. Later it became the first recognized pallasite, a group that would be named for him.

Pallasites are mixtures of crystalline olivine and a surrounding network of iron-nickel. The crystals can vary from bright yellow to light green. When the olivine is gem quality, it is called peridot. Occasionally these beautiful crystals are used in jewelry. Some are even faceted. Both the olivine and metal vary in vol.%. Sometimes the metal all but disappears leaving behind large areas of olivine with little or no metal. At other times, the silicate mineral (olivine) maintains the usual olivine to metal ratio of roughly 2 to 1. This varies considerably from specimen to specimen and even within a single specimen. If the metal is more abundant than the olivine in a given piece, the metal often develops a medium octahedrite Widmanstätten pattern, substantially increasing the beauty of the meteorite when etched. Pallasites are linked chemically, elementally and isotopically to IIIAB and IIF irons, suggesting that they may have a common origin.

The Pallasites are classified into three distinct groups or grouplets similar to the chemical groups of the iron meteorites: (1) Main Group Pallasites, (2) Eagle Station grouplet, and (3) the Pyroxene grouplet.

Main Group Pallasites

There are about 42 main group pallasites. They contain varying amounts of magnesium-rich olivine crystals in an iron-nickel matrix, usually with a volume ratio of about 2 to 1. The crystals generally have diameters of 0.5–2 cm. Main Group Pallasites are medium octahedrites (Om) with a bandwidth of 0.5–0.3. Accessory minerals such as schreibersite, troilite, and chromite are often found between the olivine and the iron. The composition of the iron-nickel is similar to the values determined for group IIIAB irons, indicating a common origin. It is thought that pallasites came from the region between the metallic core and olivine-rich mantle of a differentiated asteroid.

Eagle Station Pallasites

This grouplet was named for a pallasite that was recovered near Eagle Station, Kentucky in 1880. There are only three members, all with fragmented olivine set in an iron-nickel matrix. The olivine is very iron-rich, and the metal has a higher nickel content than members of the main group pallasites. The elemental and isotopic composition of the Eagle Station iron is similar to that of the IIF irons, suggesting that they may have originated on the same asteroid parent body. Further, isotopic evidence has suggested that there may be a connection to the CO/CV carbonaceous chondrites. The other two members of the Eagle Station Pallasite grouplet are Cold Bay and Itzawisis.

The Pyroxene Pallasites

There are only four members in this small grouplet which is characterized by minor amounts of clinopyroxene. It can appear as grains bordering the olivine, larger grains in the iron-nickel matrix or as inclusions. In elemental compositions they are similar to each other but different from the Main Group Pallasites. They are not related to any of the iron groups and may therefore represent fragments of a single previously unknown asteroid parent body. Members of this grouplet are Yamato 8451, NWA 1911, Vermillion and Zinder.

Pallasites are shown in Figures 8.1–8.8.

Figure 8.1. Krasnojarsk, the Pallas iron, type specimen for the pallasites. Weighs 282 g. Courtesy of Matt Morgan, Mile High Meteorites.

Figure 8.2. A polished face of the famous Brenham pallasite (PMG) from Kansas. These stony irons have roughly equal proportions of iron and olivine. When found, they're not pretty; terrestrial weathering severely alters the meteorite's exterior. Courtesy of Geoffrey Notkin/Aerolite.org.

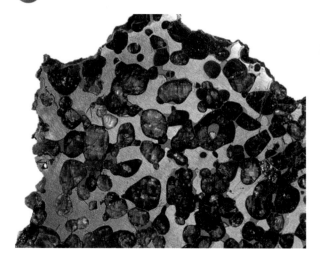

Figure 8.3. A close-up of a Brenham pallasite. The dark rim crystals (top and right) may be due to heating during atmospheric entry or to terrestrial weathering. It is estimated that the meteorites fell about 10,000 years ago. Width 11.5 cm. Photograph by Geoffrey Notkin/Aerolite.org, © The Oscar E. Monnig Meteorite Gallery.

Figure 8.4. Seymchan, a main group pallasite from Russia was found in 1967 in a riverbed. The first piece discovered had no olivine and was classified as a rare IIE iron. Subsequent finds, including this one, revealed the presence of olivine crystals, making it a pallasite. Width 8 cm.

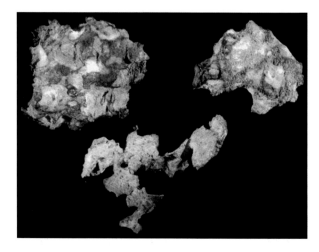

Figure 8.5. Small Imilac individuals fresh from the field. Imilac was found in the cold Atacama desert in northern Chile before 1822. Sometimes the olivine completely weathers away leaving a skeletal network of metal (bottom piece). Longest dimension of upper left piece 3.5 cm.

Figure 8.6. Esquel, a main group pallasite. It was found in Argentina 1951 by a farmer while digging a hole for a water tank. The meteorite shows beautiful yellowish green olivine (peridot) crystals. Width 6.5 cm.

Figure 8.7. The Glorieta Mountain meteorite. When cut into a thin slab, polished and lighted from behind, this becomes one of the world's most beautiful pallasites. Courtesy of Darryl Pitt, Macovich Collection, www.macovich.com.

Figure 8.8. A slab of an Eagle Station pallasite. The first Eagle Station pallasite was found in Kentucky in 1880. Iron surrounds centimeter-sized angular fragments and splinters of olivine. Impact shattered the olivine and mobilized iron, which filled the cracks between the olivine. Courtesy of Dr. Jay Piatek.

Group: Mesosiderites (MES)
Fall and find statistics: Falls—7; Finds—79
Well-known examples: Estherville, Lowicz, Vaca Muerta

Mesosiderites are typical stony-iron meteorites, most consisting of approximately equal portions of iron-nickel metal and silicates. They are polymict breccias composed of angular and rounded fragments of different mineral compositions, notably orthopyroxene, plagioclase, and eucritic material, and veins and inclusions of iron-nickel metal. The metal, which is related to the IIIAB irons, may show Widmanstätten figures. The chief problem in understanding these meteorites is that these components apparently had nothing to do with each other during their formation. They are simply accidental mixtures. The silicate fraction have compositions close to the HED portions of meteoritic igneous rock from the crust of an achondrite asteroid, and it may be that the silicates all originated on the same parent body at different levels within the crust and mantle. It is noteworthy that the most abundant mineral, olivine, which appears in virtually all chondrites is all but absent in mesosiderites. It should have occurred in large quantities in the mantle and crust of the asteroid parent.

Researchers have classified mesosiderites into groups according to mineralogy, chemical composition and the textures of the silicates. These groups may reflect origins at different depths on their parent bodies.

Mesosiderites are shown in Figures 8.9–8.12.

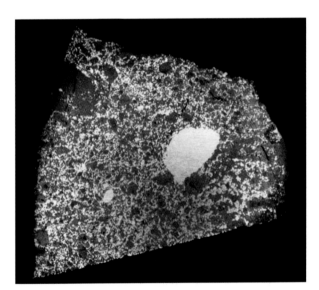

Figure 8.9. NWA 1242 (formerly Sahara 85001), a mesosiderite found in 1985 in Libya with a shock stage of 1 (S1) and weathering grade 0 (W0). It contains silicate inclusions and a 10 mm metal nodule. Width 4.5 cm.

Figure 9.5. At first glance these little spheres resemble chondrules in a chondritic meteorite. But this is spherulitic rhyolite found in Yellowstone National Park, Wyoming. These crystalline spheres grew in glassy regions of the rhyolite as the glass converted (devitrified) to tiny radiating crystals of quartz and alkali feldspar. Their diameters range from 1 to 6mm.

Figure 9.6. This L4 chondritic meteorite, called Saratov, is loaded with spherical chondrules that, in this photo, measure 0.3–2.3mm in diameter. The meteorite was observed to fall in Russia in 1914. The chondrules are easily exposed because the meteorite is fragile and crumbles readily. Courtesy of Eric Twelker, www.meteoritemarket.com.

Figure 9.7. No, this isn't a meteorite. It's bauxite, a terrestrial ore of aluminum. The aluminum oxides and hydroxides in this specimen have organized themselves into small concretions, pisoliths and ooliths during extreme weathering of soil. In this case they crudely resemble chondrules of various sizes, but none of the typical minerals (e.g., olivine and pyroxene) and textures found in chondrules can be found in bauxite. Longest dimension 9.5cm.

Figure 9.8. Well-defined chondrules dominate this meteorite as seen in this cut slab of NWA 2892 from the Sahara found in 2004. They range in size from 0.6mm to the unusually large chondrule in the center, 13mm. Note the variety of colors, shapes and rims of chondrules. Compare with the bauxite in Figure 9.7. NWA 2892 is classified as an H/L3 ordinary chondrite. Width 5cm. Courtesy of Jeff Kuyken, www.meteorites.com.au.

Figure 9.9. One of these rocks is a meteorite. Note the rounded knobby shapes in both that look like clusters of grapes. Mundrabilla (right) is an Australian iron meteorite. The knuckle-like knobs are large, randomly oriented iron-nickel crystals of taenite that stand out due to weathering. A pair of Moqui marbles (left) are concretions weathered out of Navajo Sandstone in the southwestern United States. The sand is glued together by the iron oxides, hematite and goethite. They are a terrestrial analogue to the hematite-cemented Martian blueberries seen from the Martian rover Opportunity in 2004. Long dimension of Mundrabilla 8.5 cm.

Figure 9.10. The dark brown desert varnish on this piece of granodiorite looks suspiciously like the fusion crust of a stony meteorite. But the entire light-colored interior is made of 2–6 mm crystals of plagioclase, orthoclase, biotite mica, and hornblende. This composition is not found in meteorites. Longest dimension 9.5 cm.

Figure 9.11. This round stone, found in the Franconia strewn field, was moderately attracted to a strong magnet. It lacks a fusion crust, has small holes (vesicles), and has small white spots of plagioclase. It's a cobble of water-worn terrestrial basalt with a coating of light-colored caliche where it was partially buried in the ground. Most basalt rocks are attracted to strong magnets due to the few percent of magnetite they contain. Diameter 7 cm.

Figure 9.12. This rock looks melted. In fact, it's fused soil. Hot natural or manmade fires in woodpiles often melt the soil beneath them. Little lava flows can trickle a few feet, even tens of feet, downhill from the pile. Groups of parallel lines in the upper left and upper right represent casts of carbonized rings. The exterior of a meteoroid melts in its rush through the atmosphere, but it never has casts of tree rings or large hardened bulbous drips. Width 15 cm.

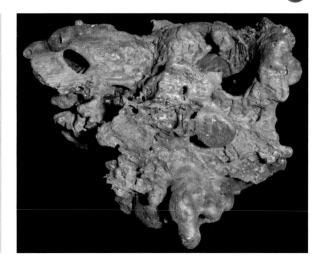

Figure 9.13. Here is a close-up of a nine-pound rock that looks and feels like an iron meteorite. It's dense and heavy, seems to have a black fusion crust, and even appears to have shallow regmaglypts. But a strong magnet is not attracted to it. Tests showed it to be nearly pure manganese, a grayish white metal (chipped area in lower middle) that oxidizes to a black color.

Figure 9.14. This bomb shell casing made of highly magnetic iron is evidence of military maneuvers. Fragments of shell casings can be torn, twisted, and rusted, but so can some iron meteorites. A test for nickel can often distinguish between nickel-free manmade iron and meteoritic iron, which always contains several percent of nickel. Longest dimension 15 cm.

Figure 9.15. The "Shirokovsky pallasite" from Russia became available in 2002, and pieces of it went into many public and private collections. But doubts about its meteoritic origin soon led to comprehensive tests and analyses. By 2004 the analyses pointed strongly to a terrestrial origin. Explanations of what terrestrial or manmade process made this meteorwrong remain controversial. The Meteoritical Society currently lists 71 pseudometeorites (high-class meteorwrongs), including Shirokovsky. Pseudometeorites are objects claimed to be meteorites, but which are nonmeteoritic in origin. Longest dimension 3.9cm.

Part III

Collecting and Analyzing Meteorites

CHAPTER TEN

In the Field

Meteorites are where you find them—all over the world. They can be anywhere on the planet for Earth is constantly being bombarded by rocks from space. Most, of course, are lost to the oceans. The vast majority are never recovered but weather away until they become completely terrestrialized. Other meteorites are trapped in glacial ice in Antarctica where they have been harvested by international teams of scientists for decades. As of 2000, more than 20,000 meteorites had been found there. None of these meteorites are available to the collector. The Antarctic Treaty was signed in 1952, became effective in 1961, and has since been ratified by more than 100 nations. In 1992, a Recommendation was adopted to protect all geological specimens, including meteorites, from private collection.

But there are plenty of places where there are no restrictions on meteorite hunting. If you are free spirited, you may want to explore remote areas. In the continental United States the best hunting ground is in the southwestern part of the Mojave desert of southern California, where vegetation is relatively sparse and the climate is dry. Look for an old surface, one that has been exposed for a long time. Old dry lakes can be a good place to search. Many meteorites have been found in Rosamond, Muroc, and Lucerne dry lakes, among others. Or you might begin by searching well-known strewn fields such as those around Glorieta Mountain in New Mexico, or Holbrook and Franconia, in Arizona. Another good location is Gold Basin, Arizona where thousands of stony meteorites have been recovered over the past decade or so (Figure 10.1).

In November 1995, Jim Kriegh (Figure 10.2), an emeritus professor from the University of Arizona, was prospecting for gold in Gold Basin, Arizona with a metal detector when he discovered two small meteorites. When Kriegh and his friends, John Blennert and Ingrid (Twink) Monrad, began hunting meteorites in earnest, they found large numbers of meteorites in the area. Most were found with a metal detector on the surface or buried up to ten inches deep in soil, and ranging from a gram to 1,500 grams. By 1998 ~2,000 meteorites had been recovered, and Dr. David Kring of the University of Arizona had established that the meteorites were L4 ordinary chondrites. By 2001, roughly 6,000 meteorites had been found by Kriegh, his friends, and numerous private collectors. Since then many more have been found and the size of the strewn field continues to grow (Figure 10.3).

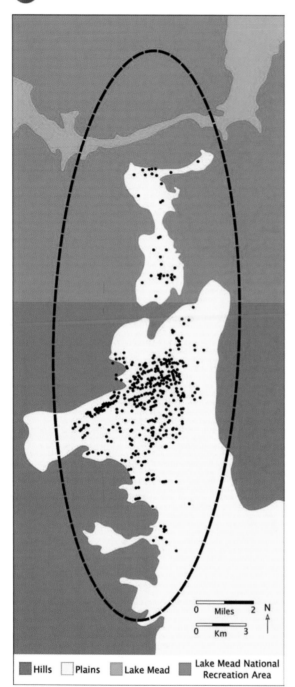

Figure 10.1. Gold Basin strewn field, the best documented strewn field on Earth. Thousands of L4 meteorites have been found following the discovery of two meteorites by Jim Kriegh in 1995. Each black dot on this map represents a single or a cluster of meteorites. The strewn field ellipse shown here will likely change shape and size as discoveries are made in the surrounding difficult, irregular, and steep terrain. The direction of the incoming meteoroid remains unknown. Typically, larger meteorites are found at the far end of a strewn field. No such size distribution has been found at Gold Basin. Data courtesy of Jim Kriegh.

Hills   Plains   Lake Mead   Lake Mead National Recreation Area

0 Miles 2   N
0 Km 3

The Gold Basin meteorites fell around 15,000 years ago in the waning years of the great Wisconsin ice age. The size of the incoming meteoroid before it broke apart in the upper atmosphere was probably 10–13 ft in diameter. After the fall, the meteorites were subjected to wetter conditions and weathered more rapidly than if they had fallen after the ice age. The Gold Basin country was

Figure 10.2. Jim Kriegh, discoverer of the Gold Basin strewn field, with one of his finds.

Figure 10.3. Howard Wells and his 119 lb collection of meteorites found at Gold Basin. Courtesy of Howard Wells.

subsequently subjected to storms and floods for thousands of years, which have washed sand, rocks, and meteorites out of the hills and onto alluvial plains. While some meteorites remain perched on bedrock and have moved little, most have been transported to new, downslope locations.

There is a vast catchment basin in the nation's wheat and corn belt extending from Texas and Oklahoma through Kansas and Nebraska. Here, after more than a century of cultivation, farmers are still clearing rocks from their fields. Any new rocks the farmers dig up have a good chance of being meteoritic. Ask permission to scout the fence rows where rocks are often thrown. More than one meteorite has been found in a farmer's rock pile, or propping open a screen door. Sometimes meteorites are found only to be lost again. In the late 1890s, Frank and Mary Kimberly home-steaded land in Kansas with a plan to clear the land for farming. But the farmers and ranchers in the Brenham Township in Kiowa County noticed curious black rocks scattered over the land. The Kimberlys noticed the rocks as well. Geologically, they didn't seem to belong there. Mary Kimberly especially took note of these rocks and saved as many as she could find. They had a curi-ous composition of iron-nickel with yellow olivine filling voids in the iron. It wasn't long before Mary Kimberly's obsession netted her well over a ton of iron rocks, much to the amusement of the neighbors. After years of collecting these black stones, the Kimberlys persuaded a geologist from Washburn College to take a look at them. He identified them as stony-iron meteorites and purchased several hundred pounds on the spot. The meteorites are called Brenham pallasites, named after the Township. The Kimberly farm near the town of Haviland was the site of the larg-est pallasite find in the United States and the strewn field lay across the farm.

Though a few pallasites continued to be found in the area through the twentieth century, it has recently experienced a flurry of activity. A 1,400-lb pallasite was dug up in a farmer's field (not only with permission, but with a lease and a lease on neighboring land as well) in October 2005 by Steve Arnold (Figures 10.4 and 10.5). It was the largest Brenham pallasite ever found. It was also

Figure 10.4. Meteorite hunters Steve Arnold and Geoff Notkin with a pallasite they excavated in the Brenham strewnfield in Kiowa County, Kansas. Pho-tograph by Sonny Clary, © Geoff Notkin.

Figure 10.5. Steve Arnold with "the big one"—the 1,400 lb Brenham oriented pallasite he found in October 2005. Photograph by Qynne Arnold, © WorldRecordMeteorite.com.

the largest oriented pallasite ever found—anywhere. He dug it up from a depth of seven-and-a-half feet, locating it with the help of a high-tech metal detector, and dug it out with a backhoe.

If you are more adventuresome you may wish to search in foreign fields. Fabulous finds of all meteorite types have been made over the last few years and continue to be made in the Sahara desert in Libya and Egypt, and in Oman (Figures 10.6–10.9). Additionally, thousands more of every kind from ordinary chondrites to lunar and martian rocks have been found in Northwest Africa. It can be dangerous to hunt in some of these locales, politically and/or physically. Check with the U.S. State Department for current restrictions and warnings. Veteran meteorite hunter Greg Hupé says, "Go with an open mind, keep you wits about you and be prepared! A lot can go wrong quick, but you can usually get yourself out of a sticky situation either through 'MacGyver-like' ingenuity to make repairs or using diplomatic reasoning to ease your way past a nervous border guard." The rewards can be great if you are willing to put up with hot deserts where you may not see another human being for days on end. Whatever you decide to do, don't forget your GPS!

You might try looking in locales where meteorites have been found in the past. Historic finds are often published in local newspapers and scientific journals and the coordinates may be known. Locating the fields takes a little detective work, but it can be well worth the time spent. *The Catalog of Meteorites* published by Cambridge University Press (2000) is a good place to start. Another is the online database of the Meteoritical Society. Both sort meteorites by country and state as well as type. In the meantime, learning as much as you can about the characteristics of meteorites is important. That's why we prepared this Field Guide.

Figure 10.6. Searching in the far reaches of the Sahara often results in stuck trucks. American collector, Jason Phillips (second from right) on expedition with Greg Hupé, December 2003. Courtesy of Greg Hupé.

Figure 10.7. Two nomads in the NWA 1068/1110 martian strewn field searching for small fragments of the meteorite. Success, a small fragment discovered in the Mars-like terrain. Note the old motorcycle fender used as a scoop. April 2002. Courtesy of Greg Hupé.

Figure 10.8. Greg Hupé (Second from left) with nomads in Berber tent in the "forbidden zone" region between the Moroccan and Algerian borders. Sometimes, travel to very remote places in the Sahara Desert is required to acquire that special stone. April 2002. Courtesy of Greg Hupé.

Figure 10.9. Sunrise over a Berber camp site in a desolate area near the "forbidden zone" between Morocco and Algeria. Beauty can be found in the harshest of environments. December 2003. Courtesy of Greg Hupé.

What Should you Look for?

When you are hunting for meteorites, remember that the appearance of a meteorite in the field depends to a large extent upon how long it has been in the terrestrial environment. Like terrestrial rocks, stony meteorites are subject to weathering, especially in wetter climates. It takes only a few years to change their appearance. So let's look both at a freshly fallen meteorite and one that has lain exposed on the surface for years. As we learned in Chap. 3, a freshly fallen stony meteorite usually has a dark brown to black crust covering it (Figure 10.10). Sometimes the fusion crust shows a fine set of fractures over the entire surface, reminding one of the crazing on a pottery glaze. This is produced by rapid cooling and shrinking of the crust. If a stony meteorite has lain on the ground for several thousand years, chemical reactions with the crust will have altered its appearance to a point where it may take on the look of the surrounding country rock (Figure 10.11). The crust lightens to a medium brown as the iron in the meteorite further oxidizes to a weathering mineral like goethite and the olivine turns to a clay-like mineral. If the meteorite remains on the surface, a shiny patina may form, giving it the appearance and composition of desert varnish.

Stony meteorites can take on many shapes, from roughly spherical to very elongated forms. Most stony meteorites are fragments, since almost all the primary bodies disrupt high in the atmosphere. This fragmentation generally dictates the final shape of the meteorite. Meteorites often break along remarkably angular lines. It is not unusual to see nearly perfect right angles between two adjacent sides with rounded edges. This distinguishes many of them from common rounded river rocks. Superimposed on these angular faces may be slight depressions or cavities ablated into the meteorite during its fiery atmospheric passage (Figure 10.12).

Figure 10.10. HaH 335, a 111 g H3 chondrite resting on the desert pavement in Libya where it was found in 2004, just a few yards from 4WD tracks. Courtesy of Dr. Svend Buhl, www.meteorite-recon.com.

Figure 10.11. Jim Kriegh, discoverer of the Gold Basin field, points to a typical Gold Basin stony meteorite which has been weathering on the desert surface for ~12,000 years.

Figure 10.12. Palo Verde Mine meteorite found in Arizona by meteorite hunter Sonny Clary. Note the remarkable angularity and shallow regmaglypts on this stone. Courtesy of Sonny Clary, www. nevadameteorites.com.

When you locate a possible meteorite, mark the location where you found it. The use of a GPS receiver is desirable here because you may have found meteorites from a strewn field where possibly dozens or hundreds of meteorites may be located. What is needed now is a fresh, unweathered face to study with a 10× hand magnifier. You might be able to file a small area on the rock with a diamond file. Or you can take the specimen to a local rock shop and ask them to cut off a small piece. The cut should be sufficiently deep to reveal an unweathered interior, usually a half inch or so. Once cut, the specimen should be carefully cleaned with 99% isopropyl alcohol and dried. Now look at the cut face again. Is the interior lighter or darker than the exterior? If it is an unweathered meteorite it will be lighter, not darker. A weathered meteorite tends to have a chemically altered interior, much like the exterior, so both may be similar. If the meteorite is an ordinary chondrite, as most are, you will immediately notice bright flakes of silvery metal scattered more or less uniformly throughout the exposed face. This is iron metal alloyed with about 5% nickel. You may see metal veins running a straight course through the cut face, or the metal may be in clusters.

Use a magnet to test the metal for magnetic attraction. The metal is not a magnet; that is, it doesn't attract other iron objects like nails but is itself attracted to a magnet. You need a strong magnet for this test, not a refrigerator magnet. (Caution: the powerful rare earth magnets can be dangerous. They are difficult to pull apart once they are joined and you can injure yourself if you don't handle them carefully. They can damage credit cards and magnetic media and they should not be used near pacemakers.) If your rock passes the test, you may have the real thing. Iron on Earth is almost always found in the oxidized state in rocks, appearing as silicates, hematite, magnetite or any number of hydrated iron minerals. This is simply because we live in an oxygen atmosphere where iron easily reacts with oxygen. Stony meteorites, on the other hand, have spent most of their existence in space; lacking oxygen, their iron remains in the elemental state. Some meteorites, the carbonaceous chondrites, for example, show clear signs of aqueous alteration in which the iron has turned to magnetite.

Having your rock pass the iron "flake test" and magnet test will get you all charged up, but remain skeptical until you look at the surface more carefully. For the next observation, you will need to do a little grinding. Prepare the surface of your suspected meteorite by sanding the cut face with silicon carbide or aluminum oxide paper, using the papers dry. Begin with #220 grit paper and proceed through #400 and #600 grit, making certain all the saw marks and pits from the preceding grades have been removed. A thick piece of glass or smooth wood upon which to attach the grinding papers works well. When you want to clean the ground surfaces, use distilled water only. Go through the various grits again, washing the specimen with distilled water until the final grade. Use distilled water for the final rinse and then submerge the specimen in 99% isopropyl alcohol. Put the washed meteorite in an oven at about 120 °F and bake the specimen for 2 or 3 hours. This will assure you that all traces of alcohol and distilled water have evaporated.

After the surface has been prepared, use your 10× magnifier and a bright light to study the texture of your specimen. Ordinary chondrites have a unique texture (see Chap. 4). If you truly have a real meteorite, you will see small, round objects like tiny marbles embedded in a fine-grained matrix. They are more or less spherical and range in diameter from about 1 to 2 mm down to 0.1 mm. These are the chondrules after which the chondrites are named. If you see these celestial marbles in your rock, you've got yourself a meteorite.

But what if you have found an achondrite, a stony meteorite without chondrules? The chances are slim that you would ever find one of the rarer meteorite types. Some have characteristics that are quite subtle and not easily distinguished from terrestrial rocks (see Chap. 5). Achondrites lack not only chondrules but also iron metal, the two most distinguishing features of the chondrites. Among these meteorites are those that resemble most terrestrial basalts. Meteoriticists believe some of them came from Mars, the Moon, and the fourth largest asteroid, 4 Vesta. Fortunately, most of the achondrites still have black fusion crusts, usually quite shiny due to a high abundance of calcium. A few have light beige to medium brown fusion crusts. It is probably best to leave the verification of these meteorites to the experts. Even they can be fooled by stony achondritic impostors.

That was the bad news. Now the good news. The irons and stony-irons are the easiest of all meteorites to distinguish in the field. In fact, so easily are the irons identified by the casual weekend rockhound that the statistics on meteorite finds are slanted heavily toward the irons. For example, of the observed falls of meteorites in which the meteorites are recovered within a few days of the fall, a mere 5.8% are irons while 93% are stones. Yet, if we tally up all the stones and all the irons worldwide including both observed falls and finds, we note that 28% of the inventory are irons. This simply means that the irons are far easier to recognize than stones, which skews the statistics. Irons are twice as heavy as ordinary chondrites per given volume, being 98% iron-nickel. There is no mistaking them (well, almost). They have exotic shapes compared to stones, and are often considered art objects by collectors and high-end auction houses.

Most of the iron meteorites found on Earth are fragments of much larger bodies that either exploded in the atmosphere or upon impact with the ground. They are quite irregular in shape, appearing mechanically bent and torn, much like exploded metal bomb casings, the debris of practice bombing runs a half-century ago.

Hunting with Metal Detectors

People often ask if they can use a metal detector to search for stony meteorites as well as irons. The answer is, you bet you can—sometimes. The relative strength of the signal, of course, depends upon the amount of metal in the meteorite as well as how deep it is buried. We saw that the amount of metal in chondritic meteorites varies enormously from as much as 20 wt.% to as little as 1 wt.%. Badly weathered meteorites will obviously result in weaker signals, since much of the elemental iron has been oxidized. In fact, completely oxidized iron meteorites will actually give a "negative" signal; that is, the signal will die out as the detector head is passed over the buried specimen. If the meteorite is freshly fallen, it will tend to stand out against the other rocks in the area. Metal detectors can be more of a hindrance in this case, serving only to slow you down. You may want to try using a "magnet cane," a strong rare earth magnet fastened to a pole (Figure 10.13).

On the other hand, metal detectors are a great help in locating specimens that are either buried completely or are among similar-colored surface rocks. This is where a metal detector comes in handy to help you "see" the surface specimen and distinguish it from the other rocks. This is especially true of old meteorites found in the southwestern deserts.

Modern metal detectors are a marvel to behold. They can easily distinguish between coins (even between denominations), gold, silver, and junk iron. The very discriminating Fisher Gold Bug 2 is a favorite with meteorite hunters and comes highly recommended (Figure 10.14).

Detectors with high discrimination should be set on the "relic" setting, since they will interpret the meteorite as an old rusty nail or other relic iron. Interestingly, at Gold Basin, Arizona, an area well-known for gold hunting, meteorites were picked up and tossed aside for years as worthless "hot rocks." In this field it is best not to try to discriminate gold from "celestial" gold, since we suspect the hunter would want both. Goldmaster 2 and 3, manufactured by White Electronics, and equivalent machines do a superb job of locating both. The signals are different but you quickly learn to distinguish between them. More advanced hunters may wish to use custom made deep seeking detectors (Figure 10.15).

Figure 10.13. Author O. Richard Norton uses a magnet cane to search for meteorites on a Nevada dry lake bed.

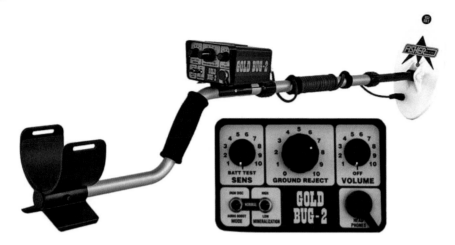

Figure 10.14. The Fisher Gold Bug-2 metal detector is easy to use for meteorite hunting. Courtesy of First Texas Products.

Figure 10.15. Greg Hupé in the Mounionalusta, Sweden strewn field holding custom-made deep-seeking metal detector loop. Note the wide oval shape to cover large areas quickly but also to walk the loop sideways between trees. Many of his finds were between or under trees. Note the field repairs using pieces of wood and duct tape. July 2007. Photograph by Devin Schrader, courtesy of Greg Hupé.

CHAPTER ELEVEN

From Hand Lens to Microscope

Using the Petrographic Microscope

For 150 years the wonderful petrographic microscope has been an essential tool for studying minerals, rocks, and meteorites (Figure 11.1). The classification of most meteorites still depends on it. Beginning in the 1960s excellent new tools became available to study rocks and meteorites such as the electron microscope, electron microprobe, neutron activation, and x-ray fluorescence. But there is no substitute for the exciting patterns and colors revealed by the venerable petrographic microscope. When rocks and meteorites are sliced thin enough, their minerals become transparent except for opaque minerals such as iron. Examining thin sections through the petrographic microscope can reveal what minerals and other building blocks are present, the pattern of their distribution, grain sizes, textures, evidence of shock from collisions, and secondary reactions such as chemical alteration and weathering. To test your skills after you have studied this chapter, you can attempt to examine and classify your meteorites using the petrographic microscope or your simpler homemade version of it.

Let's start out by following the path of light in the microscope from its source to your eye, and then consider why the petrographic microscope is unique. Light shines upward from a light source in the base of the microscope and encounters a polarizing filter, a thin section of rock or meteorite, an objective lens, another polarizing filter, and finally the eyepiece (also called the ocular lens). Note the polarizing filters (also called Polaroids or Polaroid filters). They are an essential feature of the petrographic microscope. They can be rotated to completely block light passing through them, a position called crossed polars, abbreviated XP. If a thin section of rock is placed between them, most minerals glow brightly in beautiful full colors called interference colors. Another important feature is a stage or platform that rotates allowing you to analyze changes in crystal colors, brightness, faces, cleavages, and effects of impact shock. (Many other important tools and concepts used in petrographic microscopy such as conoscopic viewing, use of the Bertrand lens, and accessory plates are not covered in this field guide.)

Figure 11.1. Author Lawrence A. Chitwood at the petrographic microscope. The microscope can reveal many of the building blocks, textures, shock, and weathering features of meteorites.

The petrographic microscope is a highly technical instrument. To fully understand and appreciate its capabilities requires considerably more knowledge than we present here. However, newcomers to the petrographic microscope will find this chapter a great help in examining meteorites.

What Is a Thin Section?

A thin section is a thin slice of rock or meteorite sandwiched and glued between a glass microscope slide and a glass cover slip (Figures 11.2 and 11.3). The process starts by sawing a rock into a small rectangular "chip," grinding one side flat, and gluing it with epoxy to a glass microscope slide. The chip is then ground down to a thickness of 0.03 mm (the thickness of fine human hair), and a glass cover slip is glued on. However, the thin section may be polished and left uncovered if it is to be examined for opaque minerals, or for chemistry with an electron microprobe or scanning electron microscope. You will notice that a standard petrographic glass slide measures 27 × 46 mm, shorter than a standard 76-mm biologic slide. When viewing a thin section in the petrographic microscope, be sure that the thin slice of meteorite and cover slip are facing up. The higher power objectives cannot get close enough to focus when the thin section is upside-down.

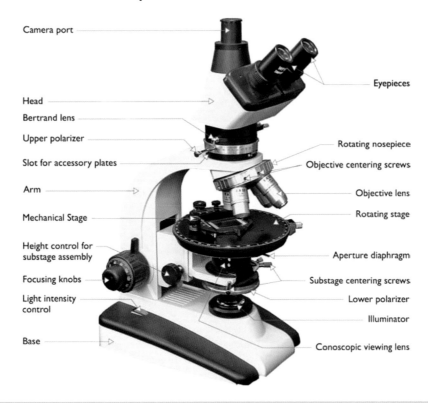

Camera port

Eyepieces

Head

Bertrand lens

Upper polarizer

Rotating nosepiece

Slot for accessory plates

Objective centering screws

Arm

Objective lens

Mechanical Stage

Rotating stage

Height control for substage assembly

Aperture diaphragm

Focusing knobs

Substage centering screws

Light intensity control

Lower polarizer

Illuminator

Base

Conoscopic viewing lens

Figure 11.9. The petrographic microscope and its parts.

describe and use the petrographic microscope. Some of the more technical features and concepts of the microscope must necessarily be ignored except where it is essential to set up the microscope (indicated by an asterisk *).

Light Source, Brightness, and Blue Filter

Turn on the *illuminator* (an incandescent light source) and, if controllable, adjust the brightness to about 90% or less of maximum. Higher settings are valuable for photography, but bulb life will be sacrificed.

Most petrographic microscopes come with a removable *blue filter*. The filter changes the yellow cast of the incandescent light to a more neutral white similar to sunlight. Place the filter over the illuminator.

Iris Diaphragm* and Conoscopic Lens*

The *iris diaphragm* and *conoscopic lens* are located immediately under the stage on the *substage assembly*. The diaphragm is an adjustable opening for light. Adjust for the full open position. A small lens called the conoscopic lens can be tilted into or out of the light beam. Be sure this lens is not in the light beam.

Polarizing Filters

The microscope has two polarizing filters, one below and one above the rotating stage. The *upper polarizer* (also called the *analyzer*) can be rotated 90 degrees or more. Normally, a scale on the rotating upper polarizer will show the amount of rotation in degrees with zero indicating the *crossed polar* (XP) position. In the zero degree position the field of view should be black.

The upper polarizer can be inserted or removed from the light path. When removed, the image of the thin section is being viewed in *plane-polarized* (PP) light.

The *lower polarizer* is normally left alone. But it can be misaligned. If the field of view is not black when the upper polarizer is set to zero degrees, rotate the lower polarizer until light is completely extinguished.

Rotating Stage

The stage of a petrographic microscope rotates a full 360 degrees and comes with a scale around its perimeter marked in degrees. Birefringent crystals in thin sections respond differently when rotated in polarized light. In cross-polarized light (XP) they brighten and dim as the stage is rotated. Every 90 degrees they turn black (or nearly black), a position called *extinction*. In plane-polarized light (PP) they may exhibit one or more subtle colors as the stage is rotated, a diagnostic feature called *pleochroism*. By rotating the stage, the position of linear features such as the edges of crystal faces and cleavages can be compared to crosshairs in an eyepiece. Whether they are parallel or oblique to the crosshairs can be diagnostic for some minerals.

Bertrand Lens*

If this lens is present, it is located above the upper polarizer. We will not use it. If it is engaged, the field of view will be much smaller and completely out of focus.

Focus Controls

Coaxial knobs adjust the height of the rotating stage and control focusing. The larger knob is the *coarse focus* and the smaller is the *fine focus*. Even though most thin sections are only 0.03 mm thick, the precise control of the fine focus knob lets you focus anywhere within the thin section. This can be useful at higher magnifications.

Objectives and Eyepieces

The lens immediately above the thin section is the *objective lens*. Most petrographic microscopes have three or four objective lenses screwed into a rotating nosepiece so that different magnifications can be easily selected. Avoid collisions between an objective lens and a thin section when the stage is raised. Thin sections and lenses can break. Objective lenses for high magnification must be very close to the thin section.

Petrographic microscopes have either one or two *eyepieces* (*oculars*). Those with two offer more refined and satisfying viewing. An eyepiece with crosshairs is preferable. If viewing with two eyepieces, crosshairs are needed for only one of them. Be sure that both eyepieces are properly focused. Typically one of them can be rotated and focused to match the other.

To determine *magnification*, look for the first number on both the objective lens and eyepiece. Multiply the two numbers together. The result is the magnification. For example, if the objective lens is labeled "5" and the eyepiece is labeled "10", the magnification is 50×. Thus, an object 0.1 mm wide will appear to be 5 mm wide.

Mechanical Stage

Many microscopes have a useful device called a mechanical stage, which can be conveniently added or removed from the rotating stage. It serves two purposes. The first is that a thin section can be carefully moved around with exquisite control by two small knobs. Also, numbers from two vernier dials can be recorded in order to relocate interesting features. The second purpose is to perform a *point count*. A point count gives the percentage of each constituent in a thin section, such as the percentage of chondrules, matrix, glass, and opaque minerals. To do a point count the thin section is moved incrementally in a grid pattern (the small knobs on most mechanical stages have dimples or gear-like teeth that allow small, precise, incremental advances of the thin section). Whatever constituent lies exactly in the crosshairs is recorded. Usually several hundred points are counted over the entire thin section. The sum of each constituent is converted to a percentage of the total number of points.

Centering Screws

Each objective on the nosepiece has a pair of centering screws. The screws are turned with a small, special tool that comes with the microscope. The purpose of this adjustment is to center each objective over the rotational center of the stage. To begin, use an eyepiece with crosshairs so you know the exact center of your field of view. Then place a thin section on the stage. While rotating the stage look for the point on the thin section around which everything rotates. Adjust the screws so that this point is exactly in the crosshairs. Center each objective beginning with the lowest magnification.

Trinocular Head

On the head of some petrographic microscopes are three viewing ports. Two are for eyepieces and the third (usually the one pointing straight up) is for a camera. Some microscopes have a sliding prism controlled by a lever that sends light to either the eyepieces or the camera. Other microscopes with a bright light source split the light beam and send it to all ports simultaneously.

Care and Cleaning

Keep the microscope as clean and dust-free as possible. Clean the lenses with great care by removing dust with compressed air and a camel hair brush. Remove oil and fingerprints with lens tissue or a clean, slightly damp, soft cloth. Avoid using solvents like alcohol; they can dissolve the cement holding the lenses together. Pick up the microscope by its base and arm only. Keep two or three extra illuminator lights handy. When not in use, cover the microscope with a plastic or cloth shroud.

Measuring the Size of Objects in Thin Sections

You can use two approaches to measuring the size of objects seen in the microscope, the estimated and the measured. In the estimated approach, look at a ruler with a millimeter scale through the microscope and estimate the diameter of the field of view. Then look at an object in a thin section and estimate its size knowing the approximate diameter of the field of view.

In the measured approach, sizes of objects seen in the microscope are measured with an eyepiece reticle scale (also called an eyepiece micrometer scale). Some eyepieces come with built-in reticle scales. Others are constructed to accept reticle scales at some later time. To accommodate different magnifications, the reticle scale must be calibrated using a stage micrometer with known distances between intervals. If you need to obtain a stage micrometer, consider one with a 2-mm-long scale with 100 intervals. Using a stage micrometer directly is impractical; the micrometer scale and objects in thin sections cannot normally be brought into focus at the same time.

Step 1. Equipment and lenses needed to begin calibration
Eyepiece with reticle scale
Lowest power objective
Stage micrometer (Figure 11.10)
Hand calculator

Step 2. Place and orient stage micrometer
Use plane-polarized light (PP) by removing upper polarizer from light path.
If necessary, rotate eyepiece until eyepiece reticle scale is horizontal.
Focus eyepiece reticle. Eyepieces with a reticle scale usually have a separate focus control on the eyepiece.
Place stage micrometer on rotating stage and move it into view (Figure 11.11). Some stage micrometers are too small to bridge the opening in the middle of the microscope stage. If so, place the stage micrometer on top of a clear part of a thin section or place it on a clear glass slide.

Step 3. Calibrating the eyepiece reticle scale
Rotate and position stage micrometer until it overlaps and is parallel to eyepiece reticle scale (Figure 11.12). At the same time, precisely position the left end of the reticle scale with the left end of the eyepiece reticle scale.
Select a point where a line on the eyepiece reticle scale overlaps a line on the stage micrometer scale. You may find overlapping lines at different points along both scales. For highest precision, select the overlap point as far to the right as possible.

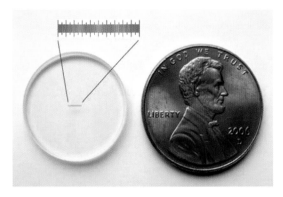

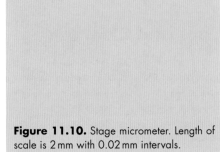

Figure 11.10. Stage micrometer. Length of scale is 2 mm with 0.02 mm intervals.

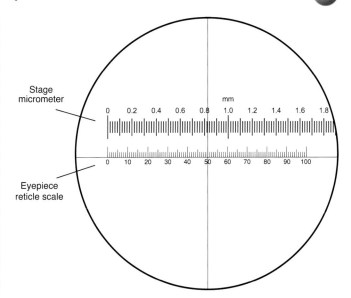

Figure 11.11. View through an eyepiece with two scales, a reticle scale mounted in the eyepiece and a stage micrometer scale placed on the microscope stage. The stage micrometer scale is used to calibrate the eyepiece reticle scale.

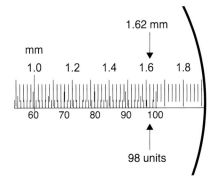

Figure 11.12. To calibrate the eyepiece reticle scale, the two scales must be overlapped and a point found where lines of each scale overlap. This point should be chosen as far up the eyepiece reticle scale as possible. To calculate the calibration factor, the stage micrometer distance is divided by the eyepiece reticle distance (1.62/98 = 0.016).

Table 11.1. Example of calibration table. The eyepiece reticle scale must be calibrated using a stage micrometer. Each objective lens must be calibrated

Objective	Stage Micrometer Distance, D	Ocular Reticle Scale Distance, N	Calibration Factor C = D/N
4X	1.48 mm	36 units	0.041 mm/unit
10X	1.62 mm	98 units	0.016 mm/unit
40X	0.40 mm	97 units	0.0041 mm/unit

Using the stage micrometer, determine the distance in millimeters (mm) from the beginning of the scale to this overlap point (D). Some micrometer scales have numbers at major divisions as in Figure 11.12, others have no numbers.

Using the eyepiece reticle scale, determine the number of units (lines) included from the beginning of the scale to the overlap point (N).

Calculate the calibration factor (C). Do this by dividing D by N; as a formula, C = D/N.

Each objective must be calibrated. Make a calibration table (Table 11.1) that shows each objective lens, distances along both scales, and the calculated distance between eyepiece scale units.

To measure the distance in a thin section, measure the distance using the eyepiece reticle scale then multiply by the correlation factor (C) for the appropriate objective lens. Keep your calibration table handy; you'll need it every time you measure distances in a thin section.

Step 4. Example

Using the steps above, the calibration table in Table 11.1 was made for a Nikon POH-3 petrographic microscope using a 2 mm stage micrometer with 50 divisions per millimeter (0.02 mm per division).

Through the 10× objective lens of the Nikon POH-3 microscope, the diameter of a chondrule measured 20 units (lines) using the eyepiece reticle scale. The calibration factor for the objective is 0.016 mm per unit (Figure 11.12). Multiply 20 by 0.0165. The diameter of the chondrule is 0.33 mm.

Using Reflected Light and Transmitted Light

Thin sections are normally viewed in transmitted light; that is, light that passes through the transparent minerals of a thin section. However, reflected light can be used to identify many opaque minerals (e.g., metal and troilite). Sophisticated petrographic microscopes are equipped with a second built-in illuminator that shines polarized light down onto the surface of a polished thin section or a small, polished sample block. Light reflected from opaque minerals carries valuable information for their identification. For our purposes, we will limit our examination to opaque minerals found in unpolished thin sections with cover slips. This limits the number of opaque minerals that can be identified, but is nevertheless valuable to see their location, shape, size, variety, and distribution.

Reflected light from polished versus ground surfaces is quite different. From polished surfaces, light reflects like a mirror. From ground surfaces, reflected light scatters in all directions because of the rough and rugged micro-landscape. The color of the reflected light is similar for both types of surfaces.

Since most petrographic microscopes do not have a built-in reflected light source, you can use an external light source to illuminate the top of a thin section. A bright flashlight with a tight beam can be used. Better yet, use a flexible fiber optic cable that conducts bright light from a lamp (Figure 11.13). At high magnifications, external illumination can be difficult because objective lenses must be very close to the thin section.

Figure 11.14 shows photos of a meteorite taken in transmitted light (PP and XP), reflected light (RL), and combinations of these. Each method reveals different valuable information.

Figure 11.13. Thin sections can be viewed using reflected light. Here, illumination is from a fiber optic cable manually aimed at the top of a thin section. A bright flashlight can also be used.

a

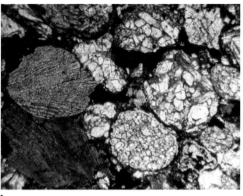

Figure 11.14. Meteorite thin sections can be examined using transmitted light or reflected light. In these examples, all photos show the same location. The type of lighting or combination of lighting reveals different information in the thin section. (Julesburg, L3.6).

a Reflected light (RL) highlights pools and stringers of iron (silvery) and troilite (bronze) between and encircling chondrules.

b

c

b Plane-polarized transmitted light (PP) reveals chondrules, their textures, and opaque areas between chondrules. Most chondrules are porphyritic. A radial pyroxene chondrule is in the upper left.

c Cross-polarized transmitted light (XP) shows the interference colors of crystals in chondrules. Brightly colored crystals are olivine and gray colored crystals are enstatite, an orthopyroxene.

d

e

d Combining reflected and plane-polarized transmitted light (RL and PP) shows chondrules, their textures, as well as areas of iron, troilite, and matrix.

e Combining reflected and cross-polarized transmitted light (RL and XP) shows iron, troilite, and matrix as well as the interference colors of crystals. Textures are less well defined.

Examining Meteorites in Thin Section

Identifying the Building Blocks of Meteorites

The petrographic microscope lets us directly see many of the building blocks of meteorites. In this section we will discuss minerals, glass, chondrules, matrix, and CAIs (calcium-aluminum inclusions) and how to identify them.

Minerals and Glass Most minerals respond to light uniquely, which means that many minerals can be identified by their combination of optical signatures. Knowledge of crystal shapes, cleavages, and other features is highly useful in identification. Seen in thin section, minerals are usually either transparent or opaque. Transparent minerals are viewed in transmitted light (light passes through the thin section). Opaque minerals are viewed in reflected light (light reflects off the top of a thin section).

Transparent minerals viewed in thin section are either isotropic or anisotropic. Isotropic minerals (such as garnet, spinel, and common table salt) remain black under crossed polars when the stage is rotated. These minerals are highly symmetric and have only one index of refraction. Light travels at the same speed in all directions. Glass (not strictly a mineral) is also isotropic. In anisotropic minerals light splits into two rays. This causes the mineral to display interference colors and to go in and out of extinction as the stage is rotated. The anisotropic minerals are divided into uniaxial and biaxial minerals. Uniaxial minerals (such as quartz, calcite, and melilite) have two indices of refraction; that is, two speeds of light. Biaxial minerals, the least symmetrical minerals, (such as olivine, pyroxene, and plagioclase) have three indices of refraction; that is, three speeds of light.

The difference between the highest and lowest indices of refraction for all anisotropic minerals is called birefringence. Birefringence varies with the orientation of a mineral in a thin section. However, for mineral identification the maximum difference (highest birefringence) is used and reported in identification guides.

Figure 11.15 shows a standard interference color chart for a standard thin section 30 μm thick. The most common minerals found in meteorites are listed next to their interference colors. Keep in mind that these are the colors produced when viewed in cross-polarized light and when the minerals are oriented in their position of highest birefringence. The color of any of these minerals can range from their highest order color (shown by the black bars) through all intermediate colors to black (at the top of the chart).

Notice that the minerals fall into three groups based on their birefringence. In chondrites, the magnesium-rich olivine crystals will display first order to high second order colors depending on their orientation. The clinopyroxenes will display up to second order colors (distinguishing between augite, diopside, and pigeonite is difficult for any petrographer). All the other minerals will display only first order colors, mostly gray and white.

But the beautiful interference colors of a mineral may not be enough to positively identify it. Other distinguishing features may be needed. Table 11.2 summarizes identifying features of the most common minerals found in meteorites. For example, the interference colors of enstatite and clinoenstatite are indistinguishable. However, the two minerals are readily identified by the absence or presence of twinning and by the position of cleavage during extinction when the microscope stage is rotated. Examples of the minerals in Table 11.2 are shown in Figures 11.16–11.26.

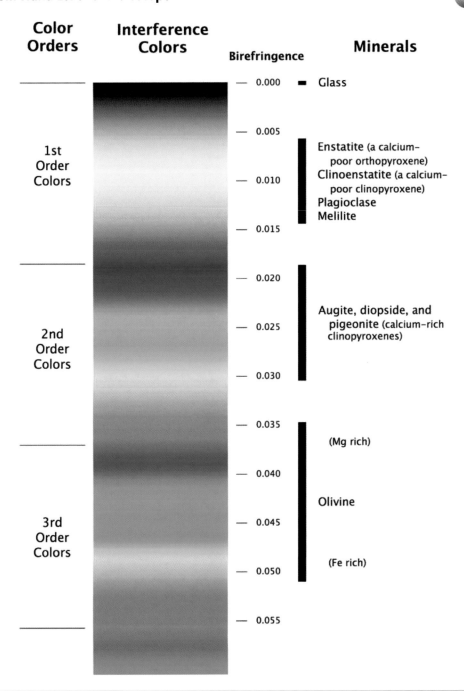

Figure 11.15. Interference color chart for thin sections 30μm thick. When a mineral is viewed in cross-polarized light, it will display an interference color corresponding to its position on the chart. However, the mineral must be oriented to present its highest birefringence. Since orientation of minerals is often random, interference colors can range from black (zero birefringence) to its maximum color (highest birefringence). The black bars represent the range of maximum birefringence for these minerals due to a range of possible chemical compositions. The first four minerals all have a similar range of birefringences.

Table 11.2. Minerals commonly found in chondrites with identifying features seen in the petrographic microscope

Mineral	Composition	Interference Colors	Twinning	Cleavage	Reflected Light
Olivine	(Mg,Fe)2SiO4	High 2nd order	No	No	–
Orthopyroxene (enstatite, a low calcium pyroxene)	(Mg,Fe)2Si2O6	Middle 1st order	No	2 cleavages; extinction on crosshairs	–
Clinopyroxene (clinoenstatite, a low calcium pyroxene)	(Mg,Fe)2Si2O6	Middle 1st order	Yes	2 cleavages; extinction oblique to crosshairs	–
Clinopyroxene (pigeonite, augite, diopside, fassaite)	Ca(Mg,Fe)Si2O6	2nd order	Yes	2 cleavages; extinction oblique to crosshairs	–
Plagioclase	Ca(Al2Si2O8)-Na(AlSi3O8)	Middle 1st order	Yes	2 oblique cleavages	–
Mellilite	(Ca,Na)2(Al,Mg)(Si,Al)2O7	Middle 1st order	No	Yes	–
Maskelynite (plagioclase converted to glass)	Ca(Al2Si2O8)-Na(AlSi3O8)	Isotropic (always extinct)	No	No	–
Glass	Wide range of compositions	Isotropic (always extinct)	No	No	–
Clay Minerals (hydrous phyllo-silicates)	Sheet silicates with water	1st and 2nd order (in fine grained aggregates)	No	Yes	–
Iron-nickel	FeNi mixture (nickel 5-20%)	Opaque	–	–	Grayish white to silver white
Troilite	FeS	Opaque	–	–	Bronze

Figure 11.16. Olivine. These stubby crystals, many with straight-line crystal face edges, lack cleavage (but have been fractured) and show a range of first and second order interference colors due to their random orientations. (Jilin, H5).

Figure 11.17. Enstatite (calcium-poor orthopyroxene). When the straight, parallel lines of cleavage are aligned with either EW or NS eyepiece crosshairs, the crystal goes into extinction (turns black). (NWA 1182, HOW).

Figure 11.18. Clinoenstatite. Enstatite and clinoenstatite look similar and display first order gray color in XP. They can be distinguished by twinning and inclined extinction, both found only in clinoenstatite. Twinning in clinoenstatite appears as parallel, thin, alternating light and dark gray bands in XP (right photo) but cannot be seen in PP (left photo). (Parnallee, LL3.6).

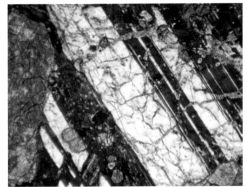

Figure 11.19. Calcium-rich clinopyroxene. Second order interference colors, two cleavages at close to 90 degrees to each other, inclined extinction, and widely spaced twinning mark this crystal as a calcium-rich clinopyroxene. (NWA 1909, EUC).

Figure 11.20. Plagioclase. Twinning with alternating dark gray and white bands of variable width is common in plagioclase. XP (Dhofar 007, EUC).

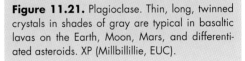

Figure 11.21. Plagioclase. Thin, long, twinned crystals in shades of gray are typical in basaltic lavas on the Earth, Moon, Mars, and differentiated asteroids. XP (Millbillillie, EUC).

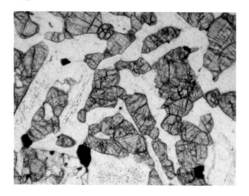 

Figure 11.22. Plagioclase and calcium-rich clinopyroxene. In plane-polarized light (left photo), white areas are plagioclase, medium brown areas are clinopyroxene, and small black areas are an opaque mineral. In cross-polarized light (right photo), plagioclase appears in various shades of gray with some twinning. Clinopyroxene displays high first order and low second order interference colors. (NWA1909, EUC).

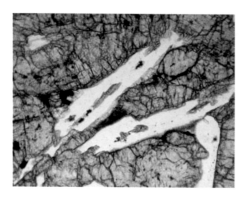 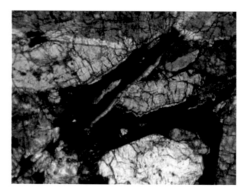

Figure 11.23. Maskelynite. In plane-polarized light (left photo), the white areas are maskelynite, a glass that chilled from shocked and melted plagioclase crystals. In cross-polarized light (right photo), the isotropic glass is black and stays black when the microscope stage is rotated. A calcium-rich clinopyroxene glows with bright second order interference colors. (Zagami, Shergotite).

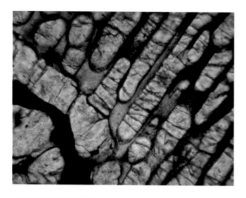 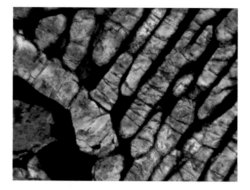

Figure 11.24. Glass in barred olivine chondrule. Compare the medium gray areas in the left photo (PP) with the corresponding black areas in the right photo (XP). These areas are glass surrounded by bars of a single dendritic olivine crystal connected in three dimensions. (Barratta, L3.8).

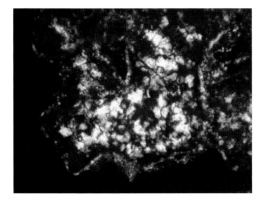

Figure 11.25. Mellilite in CAI. Found commonly in CAIs, mellilite lacks twinning and displays first order shades of gray to white. XP (Allende, CV3.2).

Figure 11.26. Silvery white material is iron-nickel; bronze material is troilite. Grainy texture of both due to abrasives used to prepare thin section. RL (Julesburg, L3.6).

Some minerals display pleochroism when viewed in plane-polarized light (PP). A mineral is pleochroic if you see one or two colors as you rotate the stage. The colors can be subtle and washed out. In most meteorites the pyroxenes are magnesium-rich pyroxenes and show little or no pleochroism. But with increasing iron content the pyroxenes become increasingly pleochroic with colors of pale yellow, green, pink, and brown.

Chondrules Chondrules are the quintessential unearthly feature of most meteorites (Figures 11.27–11.33).

Their generally spherical shape and textures have survived the vast time between the origin of our solar system and today. Seen in the petrographic microscope, their variety and beauty stand unchallenged among the building blocks of meteorites. Look over the photographs in the Chondrule Gallery of Chap. 4 for an impression of their variety, interference colors, and textures.

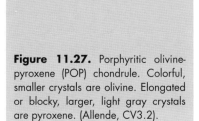

Figure 11.27. Porphyritic olivine-pyroxene (POP) chondrule. Colorful, smaller crystals are olivine. Elongated or blocky, larger, light gray crystals are pyroxene. (Allende, CV3.2).

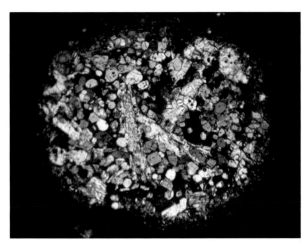

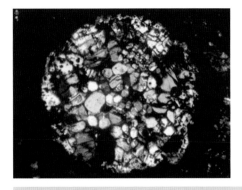

Figure 11.28. Porphyritic olivine (PO) chondrule. Virtually all crystals are olivine in various orientations. Some black and dark gray areas are olivine crystals near extinction. (Allende, CV3.2).

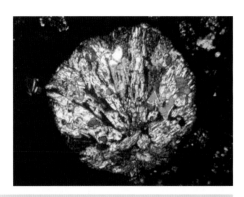

Figure 11.29. Porphyritic pyroxene (PP) chondrule. Light gray, blocky crystals of pyroxene (clinoenstatite) appear to emanate from a point in some PP chondrules. A few olivine crystals are scattered about. (Allende, CV3.2).

Figure 11.30. Radial pyroxene (RP) chondrule. Long, thin pyroxene crystals (enstatite) radiate from a point. Some RP chondrules have two or more radiating points. Thin crystals oriented vertically are in extinction and create the appearance of a dark shadow. (Marlow, L5).

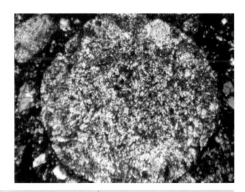

Figure 11.31. Cryptocrystalline (C) chondrule. "Crypto" refers to an abundance and confusion of tiny crystals often too small to resolve in the microscope. (NWA 096, H3.8).

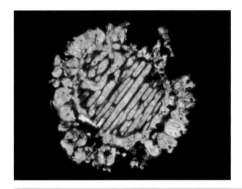

Figure 11.32. Barred olivine (BO) chondrule. Bars of olivine grew inward from a molten droplet. Dark corridors between bars are glass or altered glass. (Allende, CV3.2).

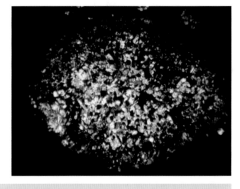

Figure 11.33. Granular olivine-pyroxene (GOP) chondrule. Small olivine and pyroxene crystals and dark glass form a mosaic. (Allende, CV3.2).

Chondrules were once molten droplets with diameters between about 0.1 and 4 mm. Thus, they are igneous with crystals that have grown in a supercooled liquid. The textures of these crystals are described as porphyritic (large crystals in a fine-grained matrix), barred (dendrites made of parallel thin blades or plates of olivine), radiating (sprays of fine-grained fibrous pyroxene), granular (aggregate of mineral grains of roughly equal size), and cryptocrystalline (tiny crystals too small to clearly recognize in a microscope). A summary of the classification of chondrules based on these textures is given in Table 11.3. All of these types can be seen in the Chondrule Gallery.

You should be able to classify most chondrules with the help of Table 11.3 and the Chondrule Gallery.

Matrix Very fine-grained material called matrix makes up the opaque regions in chondrites in which chondrules, CAIs, crystals, and various inclusions are set (Figure 11.34). Optical microscopes, including the petrographic microscope, cannot resolve particles smaller than about 0.5 μm. Since much of the matrix has sizes smaller than this, the tiny mineral grains can be resolved only with an electron microscope.

Matrices are usually made up of pulverized chondrules; fragments and aggregates of metal, troilite, olivine, pyroxene and plagioclase; and pre-solar grains such as graphite, silicon carbide, corundum, and organic compounds. The matrices of each meteorite group have different histories and different combinations of materials. Matrices remain opaque in plane- and cross-polarized light.

Table 11.3. Textural classification of chondrules

Type	Texture and Minerals	Abundance (%)
POP	Porphyritic olivine-pyroxene	48
PO	Porphyritic olivine	23
PP	Porphyritic pyroxene	10
RP	Radial pyroxene	7
C	Cryptocrystalline	5
BO	Barred olivine	4
GOP	Granular olivine-pyroxene	3

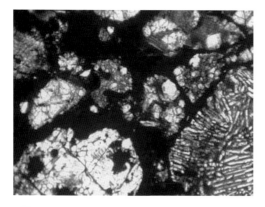 

Figure 11.34. Matrix is the black area between fragments of crystals and chondrules. In both plane-polarized light (PP, left) and cross-polarized light (XP, right) the corridors of matrix remain black indicating that glass is not a primary component of matrix. (L'hmada, LL3.5).

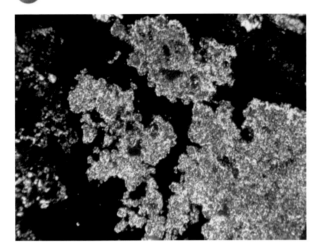

Figure 11.35. This amoeboid olivine aggregate is one of several types of CAIs (calcium aluminum inclusions) commonly found in carbonaceous meteorites, especially in CV meteorites. The bright and sometimes colored perimeter is the pyroxene diopside. (Allende, CV3.2).

However, submicroscopic matrix grains in ordinary chondrites recrystallize during thermal metamorphism. As temperatures rise, the elements of the tiny mineral grains begin to migrate and form new, larger crystals even though no melting takes place. The matrix changes from opaque to translucent to transparent. The degree of recrystallization is one of the important criteria for assigning petrographic type to ordinary chondrites.

The matrices of CI, CM, CR, and CH carbonaceous chondrites have been altered by water, with little or no alteration by temperature. The new minerals that form, such as clays (phyllosilicates), have water in their crystalline structures.

CAIs Several types of calcium aluminum inclusions (CAIs) are known, and they are found almost exclusively in carbonaceous chondrites. Many types of CAIs are well represented in CV meteorites (especially Allende). One of the most easily recognized is the amoeba-shaped CAI made of fine-grained olivine (Figure 11.35). A bright rim of diopside traces its perimeters. The origin of CAIs remains an enigma. They may have condensed as solids or liquids directly from pre-solar hot gases with enrichment from dust.

Classifying Your Chondrite

When you feel comfortable that you can identify the common building blocks of chondrites (chondrules, minerals, metal, matrix, CAIs and fragments from earlier generations of impacted parent bodies), consider testing yourself by attempting to classify known and unknown meteorites in thin section using Tables 11.4 and 11.5. Your classification may be an excellent approximation (but a defensible classification requires other essential information including chemistry).

Some chondrites are classified with a one- or two-letter symbol for a chemical group combined with a number for its petrographic type. For example, an H6 signifies a high-iron (H) ordinary chondrite strongly recrystallized (petrographic type 6); or a CV3 carbonaceous Vigarano-type chondrite that has a texture and mineralogy with little or no sign of recrystallization (petrographic type 3).

The different chemical groups of chondrites apparently came from a variety of parent bodies shattered by great impacts. Parent bodies were sites where the building blocks of meteorites accumulated in difering proportions.

Table 11.4. Classifying chondrites using distinguishing petrographic features. The four classes of chondrites—carbonaceous, ordinary, R, and enstatite—are subdivided into groups using a one- or two-letter designation

Criteria	Carbonaceous Chondrites								Ordinary Chondrites			R Chondrites	Enstatite Chondrites	
	CI	CM	CO	CV	CK	CR	CH	CB	H	L	LL	R	EL	EH
Chondrule size, mm	NA	0.3	0.15	1.0	1.0	0.7	0.02	2-10	0.3	0.7	0.9	0.4	0.6	0.2
Matrix %	>99	70	34	40	40	30-50	5	<1	10-15	10-15	10-15	36	2-15	2-15
Chondrules %	<1	20	48	45	45	50-60	70	20-40	60-80	60-80	60-80	>40	60-80	60-80
CAIs %	<1	5	13	10	10	0.5	0.1	<1	<1	<1	<1	0	<1	<1
Metal %	0	0.1	1-5	0-5	0-5	5-8	20	60-80	8	4	2	0.1	10	10
Olivine %	<1	30	>50	>50	>50	40	6	?	39	48	58	50-75	<1	<1
Pyroxene %	<1	min	min	min	min	10	60	?	28	24	16	0-13	70	65

Notes: NA = not applicable; min = minor

Table 11.5. Selected diagnostic features of petrographic types (originally called petrologic types)

Criteria	Water Alteration		Little to No Alteration	Thermal Metamorphism		
	Type 1	Type 2	Type 3	Type 4	Type 5	Type 6
Condition of chondrules	Chondrules absent	Sharply defined		Well defined	Readily defined	Poorly defined
Plagioclase	Absent			Grains 2 μm	Grains 2-50 μm	Grains >50 μm
Glass in chondrules	Absent	Mostly altered	Clear glass	Turbid	Crystalline	
Matrix	Opaque			Grades from translucent (type 4) to transparent (type 6); grain size coarsens from type 4 to type 6		
Pyroxene	Absent	Predominantly twinned		Some clinopyroxene		Orthopyroxene

Radioactivity heated some of these primitive worlds and thermally metamorphosed (recrystallized) their rocks. Chondrites carry the evidence of being heated in the form of recrystallization. For example, glass in chondrules has transformed into crystals. The petrographic types correspond to temperature and degree of recrystallization. This scale is really two overlapping scales. Meteorites of petrographic types 1 and 2 have been extensively altered by water (aqueous alteration, low temperature) and the alteration extends into type 3. Types 4–6 have been extensively altered by heat and its effects also extend into type 3. In the petrographic microscope, type 3 meteorites show little to no visible alteration. Confusing? Join the untidy world of science.

To classify your meteorite, first use Table 11.4 to determine its chemical group. Visually estimate the percentage of each listed building block from a thin section. Or for best results, use the technique of point counting to more precisely determine percentages. Secondly, use Table 11.5 to determine the petrographic type of your meteorite. Look for the features described in the criteria, and don't be surprised if some are puzzling or not applicable.

Textures—Their Look and Meaning

Textures are the distinctive or identifying features of objects or groups of objects seen in thin section such as shapes, patterns, and arrangements. They can provide us with crucial information about the origin and history of terrestrial rocks and rocks from space. The significance of some textures is still being debated. The following selected textures are briefly described and their importance discussed.

Clastic Refers to rocks made of broken fragments, or *clasts*, of older rocks. Stony meteorites can have a high fraction of angular clasts such as broken pieces of chondrules, crystals, and rocks (Figure 11.36). A *breccia* is the ultimate clastic rock made completely of sharp-angled mineral and rock fragments of all sizes including the dust-sized particles of its matrix. When the clasts and matrix of a brecciated meteorite are the same kind of rock, it is a *monomict breccia* (Figure 11.37). When the clasts have different minerals and textures, it is a *polymict breccia* (Figure 11.38). In thin section under low magnification, the clasts can be examined in detail for their similarities (monomict) and differences (polymict).

Cleavage The breaking of a mineral along its crystallographic planes due to weak atomic bonds in the crystal structure. In thin section these breaks show up as one or more sets of parallel lines (Figure 11.39). Their presence helps identify minerals. Unlike the pyroxenes, olivine has no significant cleavage.

Cryptocrystalline Crystals too small to be recognized in the petrographic microscope. Examples of this texture are cryptocrystalline chondrules and the matrix in chondrites (Figure 11.40).

Crystal Faces The terms euhedral, subhedral, and anhedral describe the degree to which crystals have well-formed faces. Euhedral crystals are completely bounded by well-formed faces (Figure 11.42). Anhedral crystals completely lack faces. Subhedral crystals have partly-formed faces. In thin section, a crystal is well-formed if its outline has symmetrical straight-line segments.

Cumulate An aggregate of touching, medium to large, subhedral to euhedral mineral grains and patches of other minerals formed by crystals that have settled to the bottom of a magma chamber. Cumulates are found on Earth and found in meteorites where they formed in large parent bodies capable of generating magma chambers.

Dendritic Refers to a mineral that has crystallized in a branching pattern similar to the feathery branches of ice crystals growing on windowpanes. Dendritic crystals often grow quickly in a fast-cooling liquid such as barred olivine in chondrules.

Embayed Refers to crystals with miniature bays (like bays for boats) where the crystal has been partially resorbed, or dissolved, back into surrounding molten rock.

Equidimensional or Equant Refers to crystals having the same or nearly the same diameter in all direction.

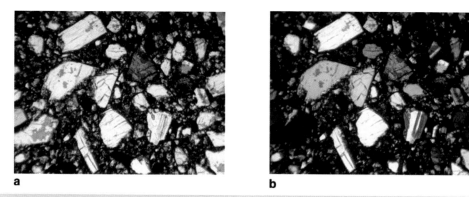

Figure 11.36. Clastic texture. This refers to broken fragments (clasts) derived from one or more preexisting rocks. In plane-polarized light (Fig. a) large and small clasts are clearly seen. In cross-polarized light (Fig. b) fewer small clasts can be seen, but interference colors assist mineral identification. (Kapoeta, Howardite).

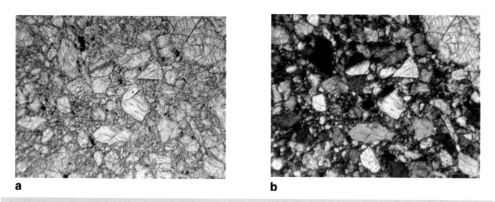

Figure 11.37. Monomict breccia. A breccia is a rock or meteorite made up of broken rock fragments. The breccia is monomict if the rock fragments have the same composition and texture. In this example, three or more rock fragments can be seen but their boundaries are indeed subtle. Mono = one or single. PP (Fig. a), XP (Fig. b). (Bilanga, Diogenite).

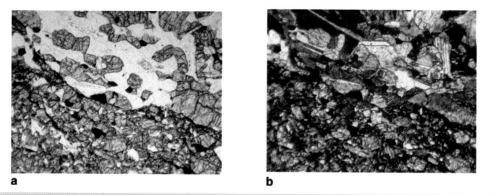

Figure 11.38. Polymict breccia. Two rock fragments are represented here with a curved boundary between them stretching from the upper left to lower right corners (seen best in plane-polarized light, left photo). In this example, crystal size and percentage of different crystals are the most distinctive differences. A polymict breccia is composed of two or more types of rock fragments. Poly = more than one. PP (Fig. a), XP (Fig. b). (NWA 1909, EUC).

Figure 11.39. Cleavage. Many crystals break along weak planes within their atomic lattices. In thin section these breaks show up as parallel lines that can be used to identify minerals. In this example of a clinopyroxene, the breaks were probably induced or enhanced by forces from a collision. PP left, XP right. (Portales Valley, H6).

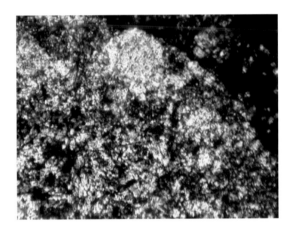

Figure 11.40. Cryptocrystalline texture. A texture with crystals too small to be resolved by optical microscopes. Crypto = hidden. XP (NWA 096, H3.8).

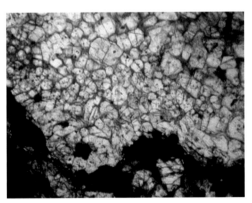

Figure 11.41. Equigranular texture. When the size of crystals in an aggregation of crystals is similar, the rock or aggregate has a equigranular texture. In this example, the crystals are essentially all olivine. PP (left), XP (right). (Moorabie, L3.8).

Equigranular (also Granular) Refers to a rock with mineral grains of roughly equal size (Figure 11.41).

Glomerocryst An aggregate of crystals of the same mineral.

Glomeroporphyritic Refers to clusters of phenocrysts scattered throughout an igneous rock.

Granoblastic Refers to a metamorphic texture of equidimensional crystals formed by recrystallization in the solid state. Some differentiated meteorites show equidimensional mineral grains crowded against one another. The grains form polygons and their sides appear to radiate outward from triple grain junctions at about 120 degrees. This is diagnostic of a meteorite that came from a region in its parent body that was highly thermally metamorphosed but not melted.

Microcrystalline Crystals small enough to be visible only under the microscope.

Ophitic An igneous texture characterized by large pyroxene grains enclosing small random plagioclase laths. Plagioclase crystals in igneous rocks are usually long and thin resembling thin strips of wood called laths. This texture is a variety of poikilitic texture.

Phenocryst A conspicuous larger crystal among smaller crystals in an igneous rock that is visible to the naked eye. The larger crystal formed earlier during a period of slower cooling.

Poikilitic An igneous texture characterized by small grains of one mineral irregularly scattered within a larger crystal of another mineral. This texture suggests that the small crystals grew first, or it can mean they both grew at the same time and the large crystal grew more rapidly (Figure 11.43).

Porous Having numerous visible or microscopic openings or pores.

Porphyritic Refers to an igneous rock in which larger crystals (phenocrysts) are set in a finer-grained groundmass.

Twinning The intergrowth of two crystals of the same mineral in a symmetrical manner. When multiple twins of the same mineral intergrow and are aligned in parallel, they are called polysynthetic twins (Figure 11.44).

Vesicular Refers to an igneous rock with abundant vesicles (holes) formed by expanding gases when the rock was molten. A type of porous rock. The meteorites Baszkówka, Seratov, and Ibitira have vesicular texture.

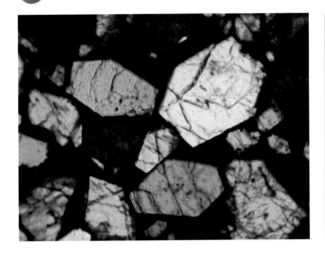

Figure 11.42. Euhedral crystals. Crystals that grow unimpeded in a melt develop characteristic shapes and faces. The faces of these coffin-shaped olivine crystals are represented by straight-line segments that circumscribe the crystals. XP (Parnallee, L3.6).

Figure 11.43. Poikilitic texture. Small grains of one mineral are irregularly scattered within a larger crystal of another mineral. In this example small colorful olivine grains are enclosed by a single large orthopyroxene crystal. XP (Mt. Tazerait, L5).

Figure 11.44. Twinning in plagioclase (left) and clinoenstatite (right). Note the refined straight twins of the plagioclase compared to the less refined, narrowly spaced twins of the clinoenstatite. Brightly colored minerals (right) are olivine. XP (left: NWA 047, EUC; right: Parnallee, L3.6).

Zoning A variation in the composition of a crystal from its core to rim. The changing composition changes the optical properties of a crystal. This is seen as a change in interference color or change in extinction angle.

The Look of Shock—Textures and Stages

Many chondrites have suffered one or more high-velocity impacts during their histories. Waves of enormous pressure pass through an impacted body inducing fracturing, crystal deformation, melting, and injection of melt into fractures. These effects are described below and used to assign a shock stage to a meteorite as shown in Table 11.5.

Undulatory Extinction A type of extinction that occurs successively in adjacent areas of the same crystal as the microscope stage is turned (Figure 11.45). Also called undulose extinction.

Mosaicism A mosaic pattern in a single crystal seen in crossed polars when the microscope stage in rotated only slightly back and forth near extinction. Shock has disorganized a crystal into separate small areas with their own different orientations and extinctions, like a set of disrupted paving stones that have been slightly tilted this way and that (Figure 11.46).

Planar Fractures Parallel sets of fractures along crystallographic planes of shocked crystals (Figure 11.47).

Shock Veins and Pockets Thin lines and sheets of glass or metal injected into veins during shock wave melting (Figure 11.48).

Maskelynite Plagioclase glass. Enormous pressure during impact converts the highly organized crystalline structure of plagioclase into a highly disordered glass. Surprisingly, the original crystal shape, faces, and edges typically survive. Twinning, a common texture of plagioclase, has been obliterated (Table 11.6).

Weathering—The Enemy of Meteorites

When meteorites fall to Earth, they find themselves on a world of abundant oxygen, water, and processes that make soil. They immediately begin to oxidize and incorporate water. Even in deserts, chemical reactions will ultimately transform all meteorites into unrecognizable soil.

A weathering scale in common use today (Table 11.7) classifies the progressive weathering of meteorites according to (1) the amount of oxidation (rustiness) of metal and troilite and (2) the alteration of silicate minerals (primarily olivine, pyroxene, and plagioclase) to clays and oxides. Metal (FeNi) and troilite are immediately and progressively affected in grades W1 through W4. Only in the rare grades of W5 and W6 are the silicates affected.

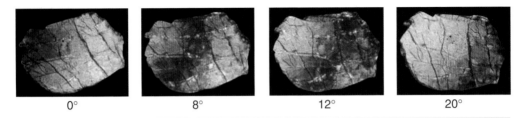

0°　　　　8°　　　　12°　　　　20°

Figure 11.45. Undulatory extinction in olivine, pyroxene, and plagioclase crystals of meteorites is common and demonstrates that the crystals endured high shock pressures in their past. A wave-like shadow sweeps across this shocked olivine crystal when the microscope stage is rotated. An unshocked crystal darkens and goes extinct uniformly when rotated. XP. (Moorabie, L3.8).

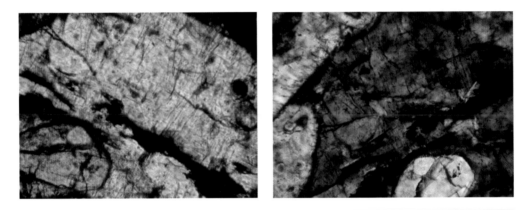

Figure 11.46. Mosaicism. The left photo shows an orthopyroxene crystal rotated to its brightest position. In the right photo, the orthopyroxene has been rotated counterclockwise to its extinction position. Instead of being fully extinct (black), a mottled pattern of dark gray colors indicates mosaicism. The green mineral is olivine. XP. (Barratta L3.8).

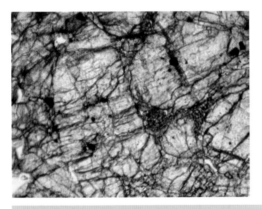

Figure 11.47. Planar fractures. Note the parallel fractures in the tall olivine immediately left of center. PP (Tenham, L6).

Figure 11.48. Shock veins form an irregular black loop in the center of the photos. A connected black vein trails off to the upper right. The veins are primarily made of glass but can alter into opaque material as in this example. PP (Barratta, L3.8).

Table 11.6. Shock classification of chondrites. A shock stage is assigned to chondrites based on the effects seen in thin section. After Stöffler, Keil, and Scott.

Shock Stage	Effect on Crystals	Local Effects	Shock Pressure (GPa)*
S1 – Not shocked	Sharp optical extinction.	None	–
S2 – Very weakly shocked	Undulose extinction. Some planar fractures in pyroxene.	None	< 4 – 5
S3 – Weakly shocked	Undulose extinction. Planar fractures in olivine. Lamellae in pyroxene.	Opaque shock veins, some melt-pockets, may be interconnected.	5 – 10
S4 – Moderately shocked	Weak mosaicism and planar fractures in olivine and pyroxene. Plagioclase partly converted to maskelynite.	Opaque shock veins, some melt-pockets and veins interconnecting.	10 – 15
S5 – Strongly shocked	Strong mosaicism and planar fractures. Plagioclase completely converted to maskelynite.	Pervasive melt pockets, veins and opaque shock veins.	25 – 30
S6 – Very strongly shocked	Melting of crystals in localized regions.	Pervasive melt pockets, veins and opaque shock veins.	45 – 60
Shock melted	Whole-rock melting to produce melt rocks and melt breccias.		75 – 90

* Ten gigapascals (GPa) equal 1.5 million pounds per square inch

Table 11.7. Classification of weathering grade. Thin sections are examined to determine the weathering grade of meteorites. Metal and troilite weather much more rapidly than the silicate minerals. The oxidation and hydration of metal and troilite imparts degrees of orange- to red-brown rustiness. After Wlotzka.

Weathering Grade	Characteristics
W0	No visible oxidation. Yellow-brown limonitic staining may be seen in transmitted light.
W1	Minor oxide rims around metal and troilite. Minor oxide veins.
W2	Moderate oxidation of metal, about 20-60% affected.
W3	Heavy oxidation of metal and troilite, 60-95% replaced.
W4	Complete (>95%) oxidation of metal and troilite, but no alteration of silicates.
W5	Olivine and pyroxene slightly altered, mainly along cracks.
W6	Massive replacement of silicates by clay minerals and oxides.

Weathering grades are assigned using thin sections (Figure 11.49 and 11.50). Using thin sections with cover slips in plane-polarized light (PP) and reflected light (RL) is usually sufficient to estimate the weathering grade of chondrites.

Photographing Thin Sections

Most of the photos of meteorite thin sections in this book were taken with a digital camera mounted on a petrographic microscope using the afocal method of photography. A few photos were taken without a microscope through a camera's close-up lens, an area of photography called macro photography. In this section are helpful guidelines and tips for choosing a digital camera

Figure 11.49. Weathering grade W0. Limonitic staining but no visible oxidation of metal or troilite. RL (Pultusk, H5).

Figure 11.50. Weathering grade W3. Moderate oxidation of metal and troilite. RL (Korra Korrabes, H3).

and for photographing thin sections. Once you have taken photos, the images are easily enhanced with image processing software. For additional information, internet web sites can be of great help. Afocal photography is commonly used with microscopes and with the telescopes of amateur astronomers.

Macro (Close-Up) Photography

Most digital cameras have a setting for taking close-up or macro photographs usually indicated by a flower. A close-up photo can include part or all of a thin section and provides a modest but useful 2–5× magnification when printed as a 4 × 6-in. picture. Figure 11.51 shows a general setup for taking close-up photos of a thin section. This and other setups that you might come up with will need diffuse illumination, polarizers, and color correction (white balance). Figure 11.52 shows a close-up photo taken with the setup in Figure 11.51.

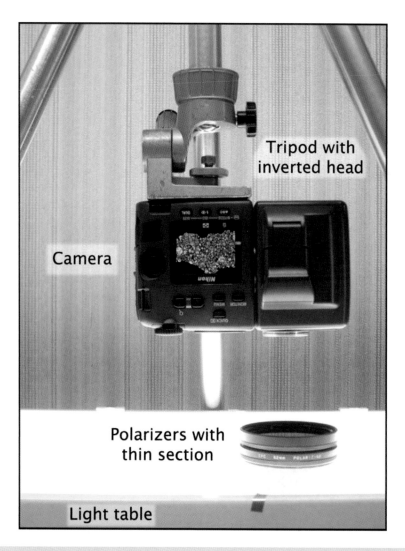

Figure 11.51. Setup for taking close-up photos of a thin section. Two polarizers with a thin section between them are placed on a diffuse source of light, in this case a light box. Arrange for the camera to be directly over the polarizers by mounting the camera on the inverted head of a tripod. Set the camera to its macro (close-up) position. Cross the polarizers so that no light passes through. Undesirable reflections from the brightly-lit camera body may make it necessary to mask light from the light table. Cut a hole for the polarizers in a sheet of black paper and lay the paper on the light table.

Figure 11.52. Example of a close-up photo of a thin section taken with the setup in Figure 11.51. This LL3 chondrite is packed with colorful chondrules and a matrix weathered to a dark brown. Image is 32 mm wide.

Diffuse Illumination A source of light must shine through the polarizers and thin section to the camera. The thin section should be illuminated evenly without bright or dark regions. You can make a simple diffuser out of a colorless, "frosted" or matte-surfaced mylar sheet used by engineers and architects. Place the sheet between the light source and the closest polarizer. In Figure 11.51 the translucent plastic cover of the light table diffuses the light from bulbs below. To highlight metal, you can use reflected light with or without transmitted light.

Polarizers You may choose to photograph in either plane-polarized light (PP) or cross-polarized light (XP). For plane-polarized light, remove the upper polarizer and place the thin section on the lower polarizer. For cross-polarized light, place the thin section between the polarizers. Be sure the polarizers are rotated to full extinction.

White Balance Sunlight provides classic white light when the sun is high in the sky and the sky is clear. If you can use it as a light source, set your camera's white balance to "sunlight" or "daylight." Since you are more likely to use incandescent illumination, you can either set your white balance to "incandescent" or better yet, create a custom white balance. In this mode, the camera will measure the color spectrum of your light source and electronically determine a unique white balance. Do this using the set up of Figure 11.51 but without a thin section and with both polarizers rotated to transmit maximum light (not crossed). Be wary of fluorescent illumination. The white balance of many digital cameras does not adequately accommodate the spiky and peculiar spectrums of most fluorescent lights.

Photography Through the Microscope

Taking photographs through a microscope with a 35-mm film camera has been routine for decades. Today, digital cameras are rapidly replacing film cameras. Small non-removable-lens digital cameras with moderate to excellent resolving power can produce quite acceptable photos through a microscope. To take photos through the microscope, place a digital camera very close but not touching the eyepiece lens, then zoom in to fill the camera's field of view (Figure 11.53). This common method is called afocal photography.

Figure 11.53. Taking photographs of thin sections through a petrographic microscope can give excellent results using the afocal method. In this method, a digital camera with automatic focus is positioned over an eyepiece and is held in place with an adaptor. This microscope has a trinocular camera port. However, one of the two eyepieces used for viewing will work just as well.

The variety of digital cameras available today can make the task of matching a camera to a microscope a challenge. The camera must be compatible with the microscope's eyepiece, and a way must be found to attach the camera to the microscope. Those cameras that accept threaded accessories have the best chance of accepting an adapter. Many companies make such adapters, which can often be found on the web. Some cameras are known to work well with microscopes including the out-of-production Nikon 990, 995, and 4500 series digital cameras. Check eBay. Adapters are readily available for these cameras (such as the ScopeTronix MaxView Plus system).

Here are several considerations to help you match and attach a camera to a microscope and to make necessary camera adjustments.

Matching Lenses The diameter of the camera lens should be roughly the same as or less than the eyepiece lens. If the camera lens diameter is too large, the photo may suffer serious vignetting (peripheral gray or black areas) that cannot be overcome by zooming.

Internal vs. External Zooming When some cameras are zoomed you can see the lens assembly move in and out. This is external zooming. A camera with this type of zoom will be more difficult to attach to and use on the microscope. We recommend using a camera with internal zoom; that is, a camera with no visible change in position of the lens assembly when zoomed.

Positioning and Attaching the Camera The camera lens should be positioned in optical alignment immediately in front of the eyepiece at a distance producing the least amount of vignetting. The distance will be on the order of millimeters to one or two centimeters. In the simplest but rather awkward case, the camera can be mounted on a tripod in front of the eyepiece. The ideal situation is to fabricate or buy an adapter that attaches the camera to the microscope's viewing port and allows the distance between the camera lens and eyepiece lens to be adjusted. Such an adapter assures ease of use and blocks extraneous light from the gap between lenses.

Zooming The digital camera should have a 3× (three times) or greater zoom capability. Use the zoom to eliminate any remaining vignetting after the camera is properly positioned. Also, use the zoom to magnify an area of interest on the thin section.

Manual Controls You will need to override certain automatic functions of the camera including setting the white balance, turning off the flash, experimenting with exposure compensation, and setting the aperture to its largest opening (lowest f-stop or f-number).

Focusing Set your camera to macro (close up) mode. This will provide the widest range of automatic focusing. When the focus of the microscope is properly set for your eyes, the camera will likely automatically focus easily. If you're unsure, you can raise and lower the microscope stage slightly with the fine focus knob to check that the camera's focusing system tracks your changes. Take a photo anywhere within the focusing range of the camera.

White Balance Correcting for artificial light is essential. Use the white balance procedure described in the previous section (Macro Photography). A blue filter usually placed over the illuminator of the microscope can remain in place while setting the white balance. However, it may alter the color spectrum of the illuminator's light too much to be corrected by the camera. To check, take two photos. Take a photo with the white balance corrected for the blue filter. Take a second photo with the white balance corrected for light without the blue filter. The second photo will be your standard. Compare the photos. If the colors of the blue filter photo are different, don't use the blue filter for photography.

Illumination Most microscope illuminators provide adjustable levels of light from incandescent bulbs. Each adjustment of the illuminator changes the color spectrum of light entering the microscope. Most digital cameras can accommodate these changes if the white balance is reset each time. Minimize changing the light level and you minimize the need to reset the white balance.

Useful Web Sites

How to make thin sections

http://almandine.geol.wwu.edu/~dave/other/thinsections

Measuring size of objects in thin sections

http://www.microscope-depot.com/ret_choose.asp

References

MacKenzie WS and Adams AE. *A Color Atlas of Rocks and Minerals in Thin Section*: Manson Publishing Ltd; 1994. 192 p.

Lauretta DS and Killgore M, *A Color Atlas of Meteorites in Thin Section*, Golden Retriever Publications and Southwest Meteorite Press; 2005.

MacKenzie WS, and Guilford C, *Atlas of rock-forming minerals in thin section*, Longman Group Limited; 1980.

MacKenzie WS, Donaldson CH, and Guilford C, *Atlas of igneous rocks and their textures*, Longman Scientific and Technical; 1982.

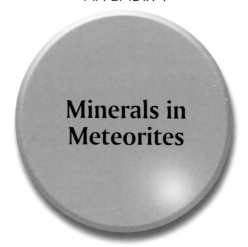

Minerals in Meteorites

Minerals make up the hard parts of our world and the Solar System. They are the building blocks of all rocks and all meteorites. Approximately 4,000 minerals have been identified so far, and of these, ~280 are found in meteorites. In 1802 only three minerals had been identified in meteorites. But beginning in the 1960s when only 40–50 minerals were known in meteorites, the discovery rate greatly increased due to impressive new analytic tools and techniques. In addition, an increasing number of different meteorites with new minerals were being discovered.

What is a mineral? The International Mineralogical Association defines a mineral as a chemical element or chemical compound that is normally crystalline and that has been formed as a result of geological process. Earth has an enormously wide range of geologic processes that have allowed nearly all the naturally occurring chemical elements to participate in making minerals. A limited range of processes and some very unearthly processes formed the minerals of meteorites in the earliest history of our solar system.

The abundance of chemical elements in the early solar system follows a general pattern: the lighter elements are most abundant, and the heavier elements are least abundant. The minerals made from these elements follow roughly the same pattern; the most abundant minerals are composed of the lighter elements.

Table A.1 shows the 18 most abundant elements in the solar system. It seems amazing that the abundant minerals of meteorites are composed of only eight or so of these elements: oxygen (O), silicon (Si), magnesium (Mg), iron (Fe), aluminum (Al), calcium (Ca), sodium (Na) and potassium (K). A large number of minerals found in small or trace amounts are made of less abundant elements such sulfur (S), chromium (Cr), phosphorus (P), carbon (C), and titanium (Ti).

In meteorites, two or three dozen minerals can be identified with a hand lens or petrographic microscope. The rest are opaque or too small and don't yield to examination under transmitted light optical microscopes. They require sophisticated equipment and techniques to identify such as reflected light microscopy, x-ray diffraction, electron microprobe analysis, and electron microscopy.

The most abundant minerals in meteorites are pyroxene, olivine, plagioclase feldspar, kamacite and taenite (an iron-nickel mixture), and small amounts of troilite, schreibersite, and cohenite. The silicate

Table A.1. The 18 most abundant elements in our solar system. Those elements shown in **bold** have combined to make the abundant minerals found in meteorites and on Earth

Element Chemical	Elemental Symbol	Abundance, No. of Atoms[a]
Hydrogen	H	24,300,000,000
Helium	He	2,343,000,000
Oxygen	**O**	14,130,000
Carbon	C	7,079,000
Magnesium	**Mg**	1,020,000
Silicon	**Si**	1,000,000
Iron	**Fe**	838,000
Sulfur	S	444,900
Aluminum	**Al**	84,100
Calcium	**Ca**	62,870
Sodium	**Na**	57,510
Nickel	Ni	47,800
Chromium	Cr	12,860
Manganese	Mn	9,168
Phosphorus	**P**	**8,373**
Chlorine	Cl	5,237
Potassium	K	3,692
Titanium	Ti	2,422

[a]Abundance of each element is compared to one million atoms of silicon. For example, for every million atoms of silicon there are 3,692 atoms of potassium.

minerals—pyroxenes, olivines, and feldspars—dominate the stony meteorites. Metals—kamacite and taenite—along with small amounts of schreibersite and cohenite dominate iron meteorites.

The following list is a short guide to selected minerals found in meteorites.

Silicates

Albite $NaAlSi_3O_8$
Albite is the sodium end member of the plagioclase solid solution series. It is very rare in meteorites. Minor amounts in SNC meteorites.

Anorthite $CaAl_2Si_2O_8$
Anorthite is the calcium end member of the plagioclase solid solution series. It is a common accessory mineral in chondrites and achondrites. It is a major mineral in eucrites and an accessory mineral in angrites. Also found in refractory inclusions in C chondrites.

Augite $Mg(Fe,Ca)Si_2O_6$
A calcium-rich clinopyroxene found in some achondrites. Accessory amounts in eucrites and nakhlites and a major pyroxene in shergottites.

Bronzite $(Mg,Fe)SiO_3$
An orthopyroxene in the solid solution series between magnesium-rich enstatite and iron-rich ferrosilite.

Bytownite $(Na,Ca)Al_2Si_2O_8$
A calcium-rich member of the plagioclase series. Is often found in eucrites along with anorthite and in small amounts in angrites.

Clinoenstatite $MgSiO_3$
A meteoritic pyroxene mineral. It is the end member of the monoclinic pyroxene series $MgSiO_3$ eSiO_3. Clinoenstatite can be recognized under the microscope by its low birefringence and polysynthetic twinning. Common in ordinary chondrites.

Clinopyroxene $(Ca,Mg,Fe)SiO_3$
Pyroxene minerals that formed in the monoclinic crystal system including clinoenstatite, pigeonite, augite, diopside, and hedenbergite.

Coesite SiO_2
A very dense polymorph of quartz produced by high shock pressures on quartz sandstone material. The product of crater-forming meteorite impacts. Coesite was first found around Meteor Crater in Arizona in 1957.

Diopside $CaMgSi_2O_6$
A calcium-magnesium-rich clinopyroxene. An end member of a solid solution series with hedenbergite as the iron member. Also occurs in E chondrites, aubrites and mesosiderites. Also found in small quantities in refractory inclusions in CM chondrites as interstellar diamonds.

Enstatite $MgSiO_3$
Enstatite is the magnesium-rich end member of the enstatite-ferrosilite solid solution series of the orthopyroxenes. It is the major mineral in all ordinary, carbonaceous, and R chondrites as well as the basaltic achondrites.

Fayalite Fe_2SiO_4
The iron end member of the olivine solid solution series. Fayalite content is diagnostic for petrographic types of ordinary chondrites. It is a major mineral in all chondrites except E chondrites.

Feldspars $(K,Na,Ca)(Si,Al)_4O_8$
A group of minerals including plagioclase and orthoclase.

Feldspathoids
These silicates are chemically similar to feldspars. The primary chemical difference is the amount of SiO_2 (SiO_2 is called silica). Feldspathoids contain about two-thirds the silica as the feldspars. The common meteoritic feldspathoids are nepheline $(Na,K)AlSiO_4$ and sodalite $Na_4(Si_3Al_3)O_{12}Cl$. They are found in chondrules and refractory inclusions in CV chondrites.

Forsterite Mg_2SiO_4
The magnesium end member of the olivine solid solution series. See fayalite.

Glass
Common in many chondrites and achondrites. Because glass has no crystalline structure it is not considered a mineral. In meteorites, glass forms when molten silicate material cools rapidly and crystals have insufficient time to grow. Glass can crystallize if heated (but not to melting) and then slowly cooled. Minerals can turn into glass from high-pressure impact (see maskelynite).

Hypersthene $(Mg,Fe)SiO_3$
An orthopyroxene of the solid solution series enstatite to ferrosilite. It is more iron-rich than either enstatite or bronzite. Hypersthene is a major phase in diogenites appearing a light green or brown. It is also common in L-group ordinary chondrites.

Maskelynite $(Na,Ca)(Si,Al)_3O_8$
Has the composition of plagioclase and has been transformed into a glass by shock metamorphism. Most commonly found in shocked plagioclase-bearing shergottites and ordinary chondrites. Presence of maskelynite glass is diagnostic for shocked meteorites that suffered impact pressures of ~30 GPa or higher.

Mellilite $(Ca,Na)_2(Al,Mg)(Si,Al)_2O_7$
A complete solid solution series with compositions ranging between akermanite, $Ca_2MgSi_2O_7$, and gehlenite, $Ca_2Al(Si,Al)_2O_7$. Found in CAIs (calcium aluminum inclusions) in CV chondrites and in large chondrules of the Allende CV3 chondrite.

Olivine $(Mg,Fe)_2SiO_4$
A complete solid solution series of minerals ranging from magnesium-rich forsterite to iron-rich fayalite. The composition of olivine is usually expressed as the molecular percentage of fayalite (e.g., Fa_{20}); the remaining percentage of forsterite is assumed. Magnesium-rich olivines are much more common in meteorites than iron-rich olivines. Olivine is a major mineral in all chondrites, pallasites, and some achondrites, but is rare in E chondrites and aubrites. See fayalite and forsterite.

Orthoclase $KAlSi_3O_8$
Very rare in meteorites. Found in accessory amounts in eucrites and in nakhlites.

Orthopyroxene $(Mg,Fe)SiO_3$
Pyroxene minerals that formed in the orthorhombic crystal system including enstatite (also called orthoenstatite), ferrosilite, bronzite, and hypersthene. They are often referred to as low-calcium pyroxenes.

Phyllosilicates
This large class of hydroxyl-bearing and water-bearing minerals usually forms in stacked flat sheets. They comprise several groups and minerals including the serpentine group, smectite (clay) group, mica group and chlorite group. Of the four, the first two groups are most important to meteorites. They occur as a result of aqueous alteration of meteoritic minerals and are found most commonly in carbonaceous chondrites.

Pigeonite $(Fe,Mg,Ca)SiO_3$
A Ca-poor clinopyroxene with 5–15 mole% $CaSiO_3$. It is a major phase in eucrites and a cumulate mineral along with augite and orthopyroxene in shergottites. Iron-rich olivine is rimmed by pigeonite in nakhlites.

Plagioclase $(Na,Ca)(Si,Al)_3O_8$
A complete solid solution series of minerals ranging from anorthite (Ca-rich) to albite (Na-rich). See anorthite and albite.

Pyroxenes
A group of minerals including orthopyroxenes (e.g., enstatite) and clinopyroxenes (e.g., augite, diopside, and pigeonite). The composition of a pyroxene is more precisely stated in terms of three end members of the pyroxene composition system $CaSiO_3$–$MgSiO_3$–$FeSiO_3$. These end members

correspond to the minerals wollastonite (Wo), enstatite (En), and ferrosilite (Fs) and are reported in terms of molecular percentages (e.g., $Wo_{42}En_{54}Fs_4$).

Quartz SiO_2
Extremely rare in meteorites. Found in small quantities in eucrites, other calcium-rich achondrites, and in the highly reduced E chondrites.

Ringwoodite $(Mg,Fe)_2SiO_4$
An olivine with a spinel structure. First found in shock veins in an ordinary chondrite in 1969. A high-pressure mineral in which magnesium-rich olivine is converted to ringwoodite at pressures of about 150 kbar or more. An indicator of impact shock in meteorites.

Serpentine $Mg_3Si_2O_5(OH)_4$
A group of hydrous minerals produced by the aqueous alteration of the magnesium silicates, olivine and pyroxene, in meteorites. Abundant in the matrices of CI and CM chondrites, usually fine-grained and mixed with organic matter.

Smectites
A group of clay minerals with complex compositions including montmorillonite and saponite. These have been found in CM chondrites and SNC (Martian) meteorites.

Stishovite SiO_2
A high-pressure extremely dense polymorph of quartz produced by meteorite impact into quartz-bearing rock. It is usually associated with coesite and forms at static pressures of over 100 kbar. Its occurrence is diagnostic for terrestrial impact craters.

Wollastonite $CaSiO_3$
End member in the pyroxene composition system $CaSiO_3$–$MgSiO_3$–$FeSiO_3$. Frequently the composition of a pyroxene is stated in terms of molecular percentages of these three end members: Wo (wollastonite), En (enstatite), and Fs (Ferrosilite).

Carbonates

Calcite $CaCO_3$
Rare in meteorites. Sometimes found along veins in CI chondrites. Often found associated with magnetite.

Hydroxides

Akaganeite β-$FeO(OH,Cl)$
This is a major corrosion product in the terrestrial weathering of FeNi in all meteorites. Akaganeite is the major carrier of chlorine indigenous to the environment, but not necessarily in meteorites. The low nickel iron, kamacite, is converted directly to akaganeite within the meteorite.

Goethite α-$FeO(OH)$
A major secondary mineral and the product of terrestrial weathering of FeNi in meteorites.

Oxides

Chromite $FeCr_2O_4$
　　Found in many meteorite groups. It is the dominant oxide in ordinary chondrites. Often found as small, black and opaque euhedral and subhedral grains in chondrules.

Ilmenite $FeTiO_3$
　　A black, opaque, slightly magnetic mineral; the principal ore of titanium. Occurs as a common accessory mineral in terrestrial igneous rocks, achondrites, lunar mare basalts and martian basalts.

Magnetite Fe_3O_4
　　Opaque, black, strongly magnetic iron oxide. Commonly found in the matrix of carbonaceous chondrites and in small amounts in ordinary chondrites and some achondrites. A common mineral in the fusion crusts of stony meteorites and forms a black coating on terrestrially weathered iron meteorites.

Perovskite $CaTiO_3$
　　A high temperature calcium-titanium oxide found in refractory inclusions (CAIs) in carbonaceous chondrites.

Spinel $MgAl_2O_4$
　　This oxide occurs in meteorites as small, usually opaque octahedrons. It is present in small amounts in chondrules, aggregates and refractory inclusions in CV chondrites.

Sulfides

Pentlandite $(Fe,Ni)_9S_8$
　　Resembles pyrrhotite in bronze color but is not magnetic until heated. Often associated with troilite inclusions in meteorites. Found in accessory amounts in the matrix and chondrules of CO, CV, CK, and CR chondrites

Pyrrhotite $Fe1-xS$
　　A magnetic iron sulfide found in meteorites that are deficient in iron with respect to sulfur. It is similar in appearance to troilite in meteorites and is an accessory mineral in CM chondrites.

Troilite FeS
　　A bronze-colored iron sulfide occurring as an accessory mineral in nearly all meteorites. It is found as nodules in iron meteorites and is often associated with graphite nodules. In chondritic meteorites it is usually found as small blebs or grains in both chondrules and matrix averaging about 6 wt.%. It differs from pyrrhotite by lacking an iron deficiency and is not magnetic.

Phosphides and Phosphates

Schreibersite $(FeNi)_3P$
　　An iron-nickel phosphide common as an accessory mineral in iron and stony-iron meteorites. Often oriented parallel to Neumann lines in kamacite plates. Silvery white when fresh and

tarnished to bronze. Often found surrounding troilite nodules. A true extraterrestrial mineral not found on Earth except in meteorites.

Whitlockite $Ca_9MgH(PO_4)_7$

An important phosphate mineral in ordinary chondrites, R chondrites and CV chondrites. Also known as merrillite.

Carbides

Cohenite $(Fe,Ni)_3C$

Iron-nickel carbide found as an accessory mineral primarily in coarse octahedrite iron meteorites. Also found as a minor mineral in Type 3 ordinary chondrites. Oxidizes a bronze color and is often associated with schreibersite. It can be distinguished from schreibersite under a petrographic microscope.

Silicon Carbide SiC

Occurs as interstellar dust grains in the Murchison CM carbonaceous chondrite and other chondrites.

Native Elements and Metals

Awaruite Ni_3Fe

A nickel-rich iron similar to taenite found in minor amounts in CV chondrites and in small amounts in CK and R chondrites.

Copper Cu

Found widely in trace amounts in ordinary chondrites and iron meteorites. Trace amounts also found in some CV chondrites. It is usually found in tiny inclusions in FeNi and troilite.

Diamond C

A polymorph of graphite produced by shock pressures during impact either in space or on Earth. Found in some meteorites with graphite nodules and in the carbonaceous matrix in ureilites. Also found in CM chondrites as interstellar diamonds.

Graphite C

A common accessory mineral in iron meteorites, ordinary chondrites and ureilites. Occurs as nodules often associated with troilite. May be the site of diamond and lonsdaleite in IA irons and ureilites. Also found in CI and CM chondrites and some E chondrites.

Kamacite α-(Fe,Ni)

An alpha-phase (low temperature) iron-nickel metal alloy containing between 4 and 7.5 wt.% nickel. Kamacite is the principal metal in irons and stony-irons, an accessory metal in ordinary chondrites, and a minor metal in some achondrites.

Lonsdaleite C

A hexagonal polymorph of diamond. Occurs in ureilites and 1AB irons. Produced by shock metamorphism of graphite on the parent body. Lonsdaleite has been artifically produced in the laboratory.

Plessite (Fe,Ni)

A fine-grained intergrowth of kamacite and taenite commonly present in octahedrites and some chondrites.

Taenite γ-(Fe,Ni)

A gamma-phase (high temperature) iron-nickel alloy with variable nickel from 27 to 65 wt.% in iron meteorites. It occurs as thin lamellae bordering kamacite plates or as intergrowths with kamacite to form plessite.

Reference

Rubin, A.E. (1997). Mineralogy of meteorite groups. *Meteoritics and Planetary Science* 32, 231–247.

Petrographic Types

Petrographic Types

Criteria	1	2	3	4	5	6
1. Homogeneity of olivine and pyroxene compositions	—	Mean deviation of pyroxene ≥5%		<5% mean deviation to uniform	Uniform ferromagnesian minerals	
2. Structural state of low-Ca pyroxene	—	Predominantly monoclinic crystals		Monoclinic cyrstals >20%	Monoclinic cyrstals <20%	Orthorhombic crystals
3. Degree of development of secondary feldspar	—	Absent		<2µm grains	<50µm grains	>50µm
4. Igneous glass in chondrules	—	Clear and isotropic glass; variable abundance		Turbid if present	Absent	
5. Metallic minerals (maximum wt% Ni)	—	Taenite absent or very minor (Ni<200 mg/g)		Kamacite and taenite present (>20%)		
6. Sulfide minerals (average Ni content)	—	>5 mg/g		<0.5%		
7. Chondrule texture	No chondrules	Very sharply defined chondrules		Well-defined chondrules	Chondrules readily distinguished	Chondrules poorly defined
8. Matrix texture	All fine-grained, opaque	Much opaque matrix	Opaque matrix	Transparent microcrystalline matrix	Recrystallized matrix	
9. Bulk carbon (wt%)	3-5%	1.5-2.8%	0.1-1.1%		<0.2%	
10. Bulk water content (wt%)	18-22%	3-11%			<2%	

The 10 criteria for establishing a chondritic meteorite's petrographic type devised by R. van Schmus and J. Wood in 1967. Researchers today recognize 7 types.

Useful Tests

Appendix 3.1

Testing for Nickel in Iron Meteorites

How often have you looked excitedly at a rock that you thought might be a meteorite? Perhaps you found a heavy, dark gray, brown or black rock covered with deep cavities. Most intriguing is that it's attracted to a magnet, at least a little. But a piece of vesicular basalt is heavy, often full of cavities, and it too is mildly attracted to a magnet. But does your rock contain nickel? All magnetic metal in meteorites is a mixture of iron and several percent nickel. If your rock contains nickel, it's probably a meteorite. Natural Earth rocks containing an iron-nickel mixture are exceedingly rare. But be aware that many manmade iron objects often contain nickel such as some nails and metal buttons on some blue jeans.

The standard test for nickel in meteorites goes back to O. C. Farrington in 1915. He used concentrated hydrochloric acid, concentrated nitric acid, concentrated ammonium hydroxide, litmus paper, and the nickel-testing chemical dimethylglyoxime (to do his test today would require protective gloves and glasses). The only ingredients here that are not dangerous are litmus paper and dimethylglyoxime. So let's modify Farrington's test with safety, availability, and cost in mind. Replace the strong acids with household vinegar and the strong ammonium hydroxide with household ammonia. The test still remains reasonably sensitive (especially if the vinegar is heated as described in method 2). Here's a list of the chemicals and equipment you'll need for two versions of this test.

Chemicals and Equipment

- Distilled white vinegar
- Household ammonia (preferably clear without soap or other cleaning agents)
- Isopropyl or ethyl alcohol (99%)
- Dimethylglyoxime
- Glass jars with tight lids (no metal)
- Three small plastic lids (e.g., water bottle lids) (for test method 1)
- White cotton swabs (for test method 1)
- Eyedropper

Buy the vinegar, ammonia, isopropyl alcohol (99%), and cotton swabs (e.g., Q-tips) from a local grocery store or super market. Dimethylglyoxime can be easily purchased from a scientific supply company. It's sold as a low hazard light tan to white powder. Order the less expensive "lab powder" grade instead of "reagent powder" grade. The price of 25 g of dimethylglyoxime is about $17 plus about $5 shipping. You'll have a lifetime supply if you buy 25 g. You can order dimethylglyoxime from several companies including *Wards Natural Science* (http://wardsci.com) and *Science Kit and Boreal Laboratories* (http://www.sciencekit.com).

Before testing, make a 1% solution of dimethylglyoxime and alcohol. In a glass jar dissolve 1 g of dimethylglyoxime in 100 g (127 ml) of alcohol at room temperature. In English units, that's 3 level teaspoons of dimethylglyoxime in 2 cup of alcohol. Shake well to dissolve the dimethylglyoxime. This may take several minutes. As long at the lid is tight this solution will last indefinitely.

Preparing Your Unknown Rock

For this test to work, your unknown rock must contain at least some elemental iron that attracts a magnet. Some of the metal must be exposed for a successful test even if you can't see it with the naked eye. Use a wire brush—preferably a wire wheel on a bench grinder or drill motor—to remove weathered or corroded surfaces. Or use sand paper or a grinding wheel to expose a clean surface on an out-of-the-way edge or corner. Don't use a metal file; it may contain nickel. By exposing a fresh surface, the acid in the vinegar can dissolve a small amount of the metal. If the iron contains nickel, both metals will dissolve in the vinegar and your test will be positive for nickel.

Running the Test—Method 1

1 Assemble the chemicals and equipment you will need (Figure A.1). Arrange the three small lids as shown in Figure A.2. Place a small amount of vinegar in the first lid, dimethylglyoxime solution in the second lid, and ammonia in the third lid. You need only enough liquid in the lids to thoroughly moisten cotton swabs. Use the lids to make sure your stock chemicals remain pure.

2 Dip a swab in the vinegar lid, rub your unknown object for 2 to 1 min, then set the swab aside. Don't let the moistened end touch anything. If possible, preheat your unknown object to body temperature or a little higher. The vinegar is much better at dissolving nickel at higher temperatures.

Household
Clear Vinegar
Dimethylglyoxime
Household
Clear Ammonia

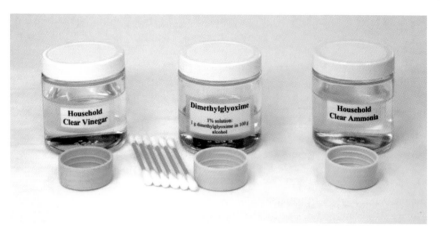

Figure A.1. Method 1 of the nickel test uses these chemicals, lids, and cotton swabs.

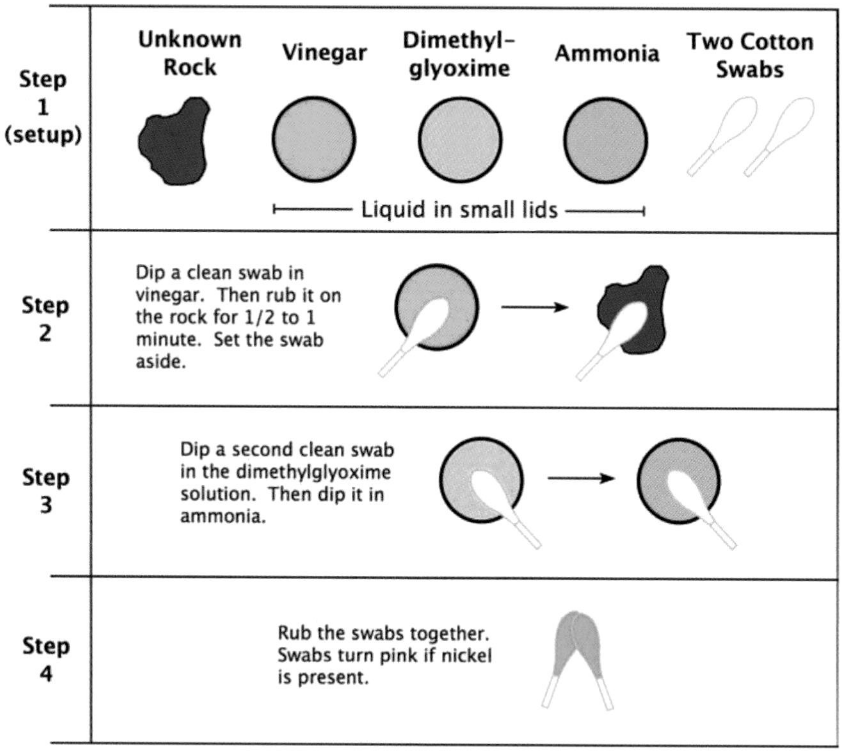

Step
1
(setup)
Unknown Rock
Vinegar
Dimethyl-glyoxime
Ammonia
Two Cotton Swabs
Liquid in small lids
Step
2
Dip a clean swab in vinegar. Then rub it on the rock for 1/2 to 1 minute. Set the swab aside.
Step
3
Dip a second clean swab in the dimethylglyoxime solution. Then dip it in ammonia.
Step
4
Rub the swabs together. Swabs turn pink if nickel is present.

Figure A.2. How to do the nickel test – Method 1.

3 Dip a second swab in the dimethylglyoxime lid, then dip this swab in the ammonia lid.

4 Rub the swabs together. If nickel is present, the swabs will turn pink. If a pink color does not appear, repeat the test and rub longer in step 2. Even a weak pink color indicates nickel.

5 Clean up by rinsing out the lids with water and then drying them. To prevent rusting and alteration, rinse the rock or meteorite with water and dry, then immerse the rock in 99% alcohol for several minutes and dry. The alcohol helps remove water and vinegar from the cracks and pores of the rock.

Running the Test—Method 2

1 Place your unknown rock in a colorless glass jar or drinking glass and add enough vinegar to cover most or all of the rock (Figure A.3). Let stand for 5–10 min. Occasionally stir the vinegar (do not use metal to stir).

To speed up this test and increase its sensitivity, warm the vinegar first (without the rock) to about 100–120 °F. Place the vinegar-containing jar in a pool of hot water from the tap, or heat in a microwave oven briefly. Place the rock in the warmed vinegar and let stand for 2 or 3 min stirring occasionally.

2 Add ammonia until the smell of vinegar changes to the smell of ammonia (the acid is neutralized), or simply add an equal amount of ammonia to the vinegar. The solution may turn yellowish brown if a lot of iron dissolved in the vinegar. Ignore this color.

3 Add a few drops of the dimethylglyoxime solution. If nickel is present, the solution will turn a vivid pink. Even a weak pink color indicates nickel.

4 Clean up by discarding the solution, rinsing the glass with water, and then drying it. To prevent rusting and alteration, rinse the rock or meteorite with water and dry, then immerse the rock in 99% alcohol for several minutes and dry. The alcohol helps remove water and vinegar from the cracks and pores of the rock.

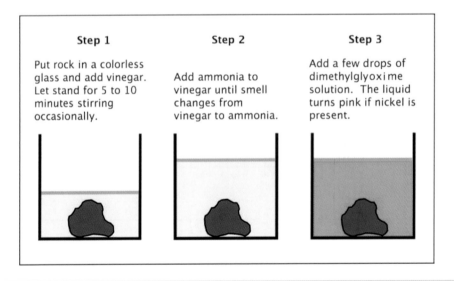

Step 1

Put rock in a colorless glass and add vinegar. Let stand for 5 to 10 minutes stirring occasionally.

Step 2

Add ammonia to vinegar until smell changes from vinegar to ammonia.

Step 3

Add a few drops of dimethylglyoxime solution. The liquid turns pink if nickel is present.

Figure A.3. How to do the nickel test – Method 2.

Appendix 3.2

Testing for Bulk Density

When we find a rock in the field we suspect is a meteorite, we immediately look for external characteristics which, if present, will indicate a true meteorite. We look for a black or dark brown fusion crust, regmaglypts or "thumb prints," and we test for magnetic attraction since most meteorites contain appreciable elemental iron. When we get it home we may slice the specimen to look for internal evidence of a meteorite. This is all well and good, but the rock may fail these tests. The fusion crust may have weathered away or been replaced by terrestrial oxides. This is what makes Gold Basin meteorites look like all the other stones in the desert pavement. Magnetic attraction works if the suspected meteorite contains elemental iron. Meteoritic iron always contains a few percent nickel, a sure sign of a meteorite.

Another useful test to distinguish some meteorites from terrestrial rocks is based on the measurement of bulk density. Most meteorites have bulk densities higher than common terrestrial rocks (Figure A.4). Thus, measuring the bulk density of a suspected meteorite can be an important step in its identification as a meteorite and in some cases what type of meteorite it is.

Using Archimedes' principle, you can easily measure bulk density with an electronic balance, water, a container to hold the water, a way to suspend a specimen, and a hand calculator. Figure A.5

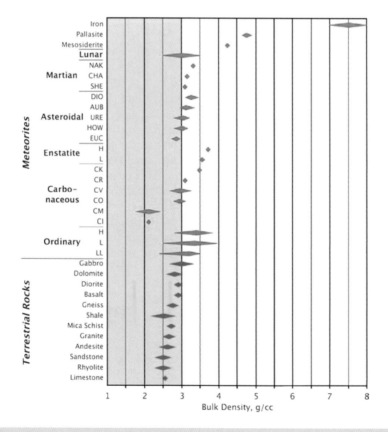

Figure A.4. Bulk density of meteorites and terrestrial rocks.

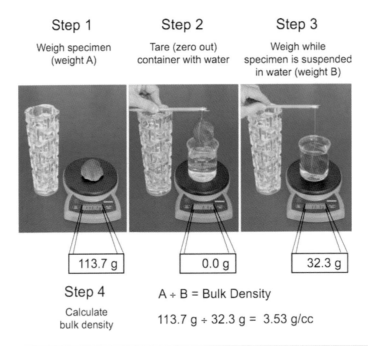

Step 1	Step 2	Step 3
Weigh specimen (weight A)	Tare (zero out) container with water	Weigh while specimen is suspended in water (weight B)
113.7 g	0.0 g	32.3 g

Step 4

Calculate bulk density

A ÷ B = Bulk Density

113.7 g ÷ 32.3 g = 3.53 g/cc

Figure A.5. How to measure bulk density using an electronic balance.

shows a step-by-step procedure for measuring bulk density. Beginning with Step 1, record the weight of the specimen using an electronic balance with an accuracy of 0.1 g. In Step 2, place a container with water on the electronic balance and tare (zero out) the balance. In Step 3, lower your specimen into the container of water using thin sewing thread tied to the specimen and to a wooden pencil. With the pencil resting on a solid support, lower the specimen into the water by rotating the pencil, similar to lowering a bucket into an old-fashioned water well with a hand crank. Make sure the specimen is completely submerged and not touching the sides or bottom of the container. Then record the weight on the balance. Finally in Step 4, calculate bulk density by dividing the weight of the specimen by the weight when the specimen is submerged.

In Step 3, the specimen may produce air bubbles while submerged. The bubbles come from air in tiny passageways in the specimen. Record the weight as soon as possible before water invades the specimen.

Figure A.5 shows an example of a specimen weighing 113.7 g and having a bulk density of 3.52 g/cc. Compare this bulk density with those of terrestrial rocks and meteorites given in Figure A.4. With a bulk density of 3.52 g/cc, the specimen is a candidate as an H or L ordinary chondrite or an L enstatite chondrite.

The vast majority of terrestrial rocks have bulk densities of 3.0 g/cc or less (the brown area of Figure A.4). Exceptions are a few rocks such as gabbro and peridotite and some terrestrial ores with bulk densities well above 3.0 g/cc. Most meteorites have bulk densities higher than 3.0 g/cc because of the elemental iron they contain. With time, meteorites lose bulk density due to weathering. Their iron and their minerals change to lower density minerals by oxidizing and hydrating. Eventually, highly weathered meteorites become indistinguishable from terrestrial rocks.

Etching Iron Meteorites

Acquiring an iron meteorite for your collection is just the beginning of a lengthy process of preparation, display and storage. Etching a meteorite is as much an artistic endeavor as it is a carefully controlled scientific experiment. Witnessing the etching of an iron meteorite is a fascinating experience. It is comparable to watching a picture appear on a piece of black and white photographic print paper during chemical development. Etching is the final process that culminates in a lengthy and carefully executed preparation of the specimen. The internal structure of an iron meteorite is best revealed by cutting and polishing a slab, especially when we wish to study the Widmanstätten patterns in octahedrites and the Neumann lines in hexahedrites. Cutting an iron-nickel meteorite requires a great deal of time and effort, often progressing no more than about an inch per hour in the cutting phase alone. Compare this to cutting a stony meteorite (ordinary chondrite) at the rate of about 10 in./h. Once the specimen has been cut to size, it is often necessary to continue with the coarse grinding phase to eliminate any deep scratches that invariably appeared on the face of the meteorite after the initial grinding.

The cutting process begins with the selection of a suitable specimen. In this case, we selected a specimen cut from a large slab of the well-known fine octahedrite, Gibeon. The finished meteorite slab should be about 5 mm thick with the front and back sides flat and as parallel as possible. Using a smooth of piece of wood about 1 in. thick as the lower grinding table, attach silicon carbide grinding paper to the edge of the wood with staples. Grind the meteorite against the abrasive, utilizing in succession these three standard grades: (#220, #400, and #600). The first grade (#220) acts to reduce or eliminate any scratches on the face of the slab that might have been introduced during the cutting of the meteorite. Using the next two grades (#400 and #600) will produce a semi-polish that is more than adequate for etching purposes.

There have been a number of chemicals used by amateurs for the etching process. The most popular uses *nitol*, a concentrated nitric acid solution mixed with 99% ethanol. When mixing the ethanol/nitric acid solution great care should be taken to see that the nitric acid is always poured *into* the *alcohol* beaker, never the reverse. This prevents splattering of the acid as it is applied to the working solution. *(Working with concentrated nitric acid is dangerous and requires handling with extreme care.)* The chemistry is simple, requiring only 15 ml of concentrated nitric acid to be mixed with about 90 ml of 99% ethyl alcohol (ethanol).

   This recipe makes 100 ml of etching solution, more than you would need for use with a single specimen. Unless you are etching many specimens at one time, you need to prepare only 1/10 of this amount. It is a good idea to wear latex gloves and protective goggles at this stage of the work. Often the specimen will be irregular in shape and does not lie flat on the work surface. It can be leveled by firmly applying a wad of clean modeling clay to back of the specimen. The slice should be placed in a shallow dish to catch the used nitol solution. Next, take a small flat paint brush approximately 10 mm wide and spread the nitol evenly over the surface with a sweeping motion. The constant motion of the brush over the surface will maintain an even flow of nitol. Within the first two minutes at this acid concentration the Widmanstätten figures will begin to appear. (Refer to the photographs in Figures A.4–A.6 to help judge the depth of the etch). As the etching proceeds, the low-nickel kamacite will slowly dissolve, leaving a bright nickel-rich taenite border. As you tilt the specimen look for the appearance of distinct kamacite plates which will appear alternately bright and dark depending upon your angle of view. When you have reached the point in your etching when the taenite looks silvery and the kamacite appears with a satin-like finish, feel free to remove the specimen from the nitol solution, and place it under slowly running tap water. At this point if you wish to darken the plates even further you can continue the etching for an additional 3 or 4 min, but usually no more.

   Nitol has been used to etch iron meteorites for many years, but recently a new etchant has appeared on the scene that is much faster acting and seems to be replacing the nitol solution as the etchant of choice. This is the PC board etchant ferric chloride ($FeCl_3$) sold by Radio Shack electronics stores nationwide. Interestingly, it is not the ferric chloride that does the etching. Rather, it is the hydrochloric acid byproduct. The acid is stronger than the nitric acid etchant and it produces a deep etch in a shorter period of time, often in less than one minute, especially if the solution has been heated to about 100 °F. Just how permanent the results will be remains to be seen. Experiments with ferric chloride etchant, compared to a nitric acid/alcohol etch, have shown that the ferric chloride produces sharper Widmanstätten figures with substantially more contrast. Moreover, many of the kamacite plates show stronger Neumann lines than normal. PC board etchant is normally used by electronic industries to dissolve unprotected copper in circuit board design. It is available at most Radio Shack electronic stores (but not online) in 16 fl oz plastic bottles. Use full strength from the bottle.

Figure A.6. A prepared surface of a 46.7 g slab of Gibeon iron before etching.

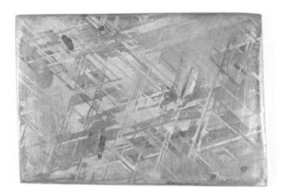

Figure A.7. A partially etched surface. The Widmanstätten figures emerge as the etchant preferentially dissolves kamacite faster than taenite.

Figure A.8. The fully etched Gibeon iron. The pattern of the three sets of kamacite bands will look different when light shines from different directions.

Unit Conversions

Length

   1 in. = 2.54 cm
   1 ft = 0.3048 m
   1 mi = 1.609 km
   1 mi = 5280 ft
   1 cm = 0.3937 in.
   1 m = 3.281 ft
   1 km = 0.6215 mi
   1 astronomical unit = 93,000,000 mi
   1 astronomical unit = 150,000,000 km
   1 light-year = 5.88 trillion mi
   1 light-year = 9.46 trillion km
   1 parsec = 3.26 light-years

Mass and Weight

   1 kg = 2.2 lb (at sea level)
   1 lb = 0.45 kg (at sea level)

Speed

   1 mi/h = 1.609 km/h
   1 km/h = 0.6215 mi/h

Pressure

    1 Pa = 0.000145 lb/in.^{2}

    1 GPa = 145,000 lb/in.^{2}

    1 lb/in.^{2} = 6897 Pa

    1 atm = 1.013 bar = 14.70 lb/in.^{2}

Volume

    1 liter = 1000 cm^3

    1 liter = 54.21 in.^{3}

    1 liter = 1.057 qt

Composition Percentages

Throughout this book we often express the amount of something as a percent of the total. The original measurements of these amounts were expressed in units of volume, weight, or moles, so the percentages are notated as vol.%, wt.%, or mole%. We report these various unit percentages as they are found in the scientific literature. They each have a specific technical meaning and a history of traditional use.

Volume percent (vol.%) refers to the volume of a constituent, such as chondrules, compared to the volume of all constituents. Point counts of thin sections give volume percents of constituents.

Weight percent (wt.%) refers to the weight (or mass) of a constituent, such as iron, compared to the weight of all constituents. In the days of wet chemistry, minerals and rocks were dissolved in strong acids and each element precipitated out as an oxide. Each precipitate was weighed and expressed as a percentage of the total weight of all precipitates (for example, SiO_2 49 wt.% or FeO 8.0 wt.%).

Mole percent (mole%) is a way to compare the number of different kinds of molecules (or atoms) in a mineral or rock. For example, the mineral olivine can be pure forsterite (Mg_2SiO_4), pure fayalite (Fe_2SiO_4), or any combination of the two (iron and magnesium are interchangeable). If we say that the olivine is 22 mole% fayalite (also written Fa_{22}), it means that for every 22 iron atoms there are 78 magnesium atoms ($22 + 78 = 100$).

See Figure A.9 for a comparison of the calculated differences of composition percentages for an imaginary meteorite made up of olivine and iron.

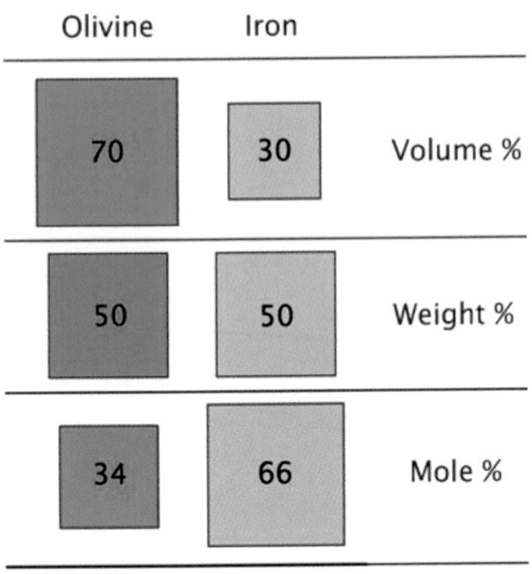

Olivine	Iron	
70	30	Volume %
50	50	Weight %
34	66	Mole %

Figure A.9. Equivalent percentages by volume, weight, and mole for an imaginary meteorite composed of 70 wt.% olivine (forsterite) and 30 wt.% iron metal.

Equipment, Storage, and Display

Meteorite collectors can benefit from a number of useful items to enhance the study, storage and display of their collections. Figure A.10 shows a number of items for a "home lab," and Figures A.11 and A.12 show a variety of items for storage and display.

Selected "home lab" items in Figure A.10:

A – Small powerful magnets. Rare earth magnets are recommended.

B – Hand lenses. 7× to 14× are best.

C – Electronic balance. Suggested minimum capacity 200 g, accuracy of 0.1 g. Preferred over beam balance because of small size and ability to "zero out" (tare) any additional weight.

D – Mechanical beam balance. Suggested capacity several hundred grams, accuracy of 0.1 g.

E – Binocular microscope. Best to have two or more magnifications or zoom magnification between 10× and 100×.

F – Thin sections and thin section boxes (see Chap. 11).

G – Petrographic microscope (see Chap. 11).

Other useful items not shown are those needed for the nickel test (Appendix 3) and for etching iron meteorites (Appendix 4).

Examples of storage and display boxes in Figure A.11:

H, I, J – Riker boxes come in various sizes with a glass window and padding. Ideal for slabs and small, uncut meteorites. Similar boxes with plastic windows are available.

K, L, M – Membrane boxes of various sizes hold specimens suspended between two thin highly elastic and tough transparent polyurethane membranes.

N – Small cardboard boxes of various sizes come with padding and opaque lids.

Specimen holders in Figure A.12:

O thru S – A sampling of various elegant brass or brass-colored specimen holders useful for creating beautiful displays of slabs and uncut meteorites.

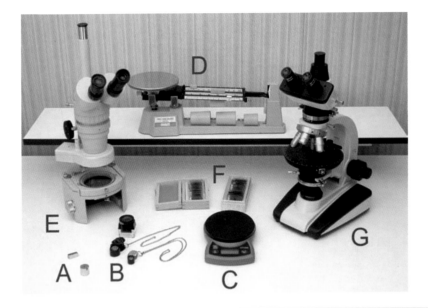

Figure A.10. A selection of useful items for a meteorite "home lab."

Figure A.11. Examples of boxes useful for storing and displaying meteorites.

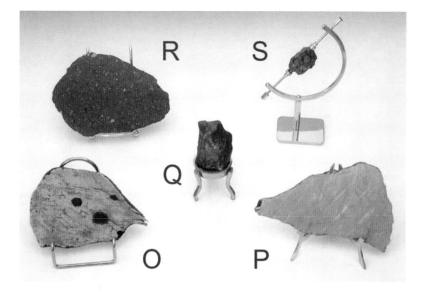

Figure A.12. Examples of elegant specimen holders for displaying slabs and uncut meteorites.

Useful Web Sites

www.jensenmeteorites.com/supplies.htm
www.membranebox.com
www.meteoritelabels.com/main.html
www.meteoritemarket.com
www.migacorp.com/meteorite_display.htm

Glossary

1 Ceres
1 Ceres is the largest known asteroid; and the first discovered in 1801.

Ablation
The removal and loss of meteoritic material by heating and vaporization as the meteoroid passes through Earth's atmosphere.

Acapulcoite
A primitive achondrite in which only partial melting and differentiation has taken place on the parent body. It has chondritic composition with some chondritic textures surviving.

Accretion
The gradual accumulation of material through the collision of particles within the Solar Nebula, or the process of interplanetary dust particles sticking together to form larger bodies.

Achondrite
A class of stony meteorites formed by igneous processes. A meteorite whose parent body has gone through melting and differentiation. These meteorites have crystallized from a magma. Achondrites include all stony meteorite types except ordinary, carbonaceous, and enstatite chondrites.

Albedo
The percentage of incoming incident light reflecting off the surface of a planetary body.

Amor asteroid
An asteroid whose perihelion distance lies just outside Earth's orbit or between 1.017 and 1.3 AU from the Sun.

Angrite
An achondritic meteorite composed of calcium, aluminum, and titanium-rich pyroxene. Accessory minerals include calcium-rich olivine and anorthite.

Anhedral
An individual mineral crystal usually of igneous origin that has failed to develop bounding crystal faces expressing its internal crystal structure.

Aphelion
In an elliptical orbit around the Sun, it is the point in the orbit where a planet is furthest from the Sun.

Apollo asteroid
A class of near-Earth asteroids defined as having a mean distance from the Sun greater than 1.0 AU and a perihelion of less than 1.017 AU. Apollo asteroids are Earth-crossers and candidate meteorite producers.

Asteroid belt
A region between the orbits of Mars and Jupiter, between 2.2 and 4.0 AU from the Sun where the Main Belt Asteroids are located.

Asteroid
A rocky or metallic orbiting body of subplanetary size showing no cometary activity; usually but not necessarily confined to the main asteroid belt.

Astronomical Unit (AU)
Mean distance between Earth and Sun; 1.496×10^8 km.

Ataxite
An iron meteorite composed of almost pure taenite with a nickel content greater than 16 wt% and showing no macroscopic structure.

Aten asteroids
An asteroid defined as having an aphelion distance of greater than 1 AU and a semi-major axis of less than 1 AU.

Aubrite
A stony meteorite formed by igneous processes containing enstatite as its primary mineral. It is also called an enstatite achondrite.

Basalt
A common fine-grained, mafic volcanic igneous rock usually erupted onto the Earth's surface from a vent or fissure. The mineral content is primarily plagioclase and pyroxene.

Basaltic achondrites
Achondrites are members of the HED class of meteorites. They have textures and compositions similar to terrestrial basalts and are believed to originate on the asteroid 4 Vesta.

Bolide
A very large meteor which is sometimes accompanied by loud sonic booms.

Brachinite
A rare primitive achondrite composed almost entirely of equigranular olivine.

Breccia
A rock made up of angular clasts of previous generations of rock cemented together by fine grained matrix material. A breccia is a common textural feature of stony meteorites.

CAI
Highly refractory inclusion rich in calcium, aluminum, and titanium. They are thought to be among the first minerals to condense out of the solar nebula. They are commonly found in C chondrites, especially CM2 and CV3 chondrites.

CB carbonaceous chondrite
A group of carbonaceous chondrites named for the type specimen, Bencubbin.

CH carbonaceous chondrite
A group of carbonaceous chondrites. The type specimen is ALH 85085.

Chassignite
A meteorite from Mars, one of the SNC group. It is similar to terrestrial dunite and made mostly of olivine.

Chondrite
A primitive stony meteorite containing chondrules. Chondrites are primitive aggregates of these chondrules.

Chondrule
Chondrules are small spherical or subspherical rock masses usually less than 1 mm in diameter which formed from molten or partly molten droplets while in space.

CI carbonaceous chondrite
A group of carbonaceous chondrites named for the type specimen, Ivuna.

CM carbonaceous chondrite
A group of carbonaceous chondrites named for the type specimen, Mighei.

CO carbonaceous chondrite
A group of carbonaceous chondrites named for the type specimen, Ornans.

Cohenite
An accessory mineral found in iron meteorites; an iron-nickel carbide. $(Fe,Ni,Co)_3C$.

Coma
The glowing part of a cometary body surrounding the nucleus and caused by radiation from the Sun.

Comet
A body that orbits the Sun and is primarily composed of frozen water, ammonia, methane and carbon dioxide along with countless pieces of rock and dust. A tail forms when it approaches the Sun from its suspected origin at the outer edge of the Solar System.

Commensurate orbit
Asteroid orbits whose periods are simple multiples or fractions of Jupiter's orbital period.

Cosmic dust or interplanetary dust particles (IDPs)
A general term for microscopic particles produced by comets as they loose their volatiles and trapped dust; iron particles may be produced by the tails of comets or by collision between asteroidal bodies; or dust shed by massive red giant stars.

CR carbonaceous chondrite
A group of carbonaceous chondrites named for the type specimen, Renazzo.

Cumulate
An igneous rock made up of relatively large crystals that settle out of a magma by gravity and accumulate on the floor of a magma chamber.

CV carbonaceous chondrite
A group of carbonaceous chondrites named for the type specimen, Vigarano.

Differentiation
A process in which a homogenous planetary body melts and gravitationally separates into layers of different density and composition. The body often separates into a core, mantle and crust.

Diogenite
An achondritic meteorite composed of magnesium-rich orthopyroxene cumulate. Diogenites may represent the upper mantle or lower crust of the asteroid 4 Vesta. It is related to howardites and eucrites that make up the HED series of achondritic meteorites.

Distribution ellipse
An elliptical area usually covering several square miles over which meteorites of a multiple fall tend to scatter. The more massive meteorites are distributed on the far end of the ellipse. See *strewn field*.

E chondrites
Enstatite chondrites; a highly reduced chondritic meteorite composed of the magnesium-rich orthopyroxene, enstatite, and iron-nickel metal.

Ecliptic
The plane of the Solar System defined as a projection of the Earth's orbit against the sky. It is the apparent yearly path of the Sun. Solar System objects are measured with respect to the ecliptic as a reference plane.

Entry velocity
A meteoroid's velocity at the beginning of the visible trail of a fireball. The initial velocity of a meteoroid at the top of the Earth's atmosphere.

Eucrite
The most common achondritic meteorite type. It is igneous in origin and is similar in composition and texture to terrestrial basalts. Possibly represents a lava flow on the surface of asteroid 4 Vesta.

Euhedral
Mineral crystals fully bounded by well-formed typical crystal faces.

Fall
A meteorite which is seen to fall and later recovered.

Fayalite
An iron -rich olivine (Fe_2SiO_4); the iron end-member of olivine.

Feldspar
A general term for aluminum silicate minerals with various amounts of sodium, calcium, and potassium.

Find
A meteorite that was not seen to fall but found at some later date. For example, many finds from Antarctica are 10,000 to 700,000 years old.

Forsterite
A magnesium-rich olivine (Mg_2SiO_4); the magnesium end member of olivine.

Fractional crystallization
A process in which, at specific temperatures, minerals crystallize out of a magma so that there are no longer reactions between the crystals and the original liquid, thereby changing the composition of the magma.

Fusion crust
A dark glassy coating that forms on stony meteorites as they ablate in the atmosphere; usually made of glass and iron oxide.

Gardening
Reworking and mixing of a regolith surface by impact from meteorites and asteroids.

Gegenschein
A faint, diffuse glowing region situated on the ecliptic plane opposite the Sun produced by sunlight reflecting off interplanetary dust particles.

Genomict breccia
A brecciated meteorite in which the individual clasts are compositionally of the same group but have differing petrographic characteristics.

Glass
A solid material that has no crystal structure. Glasses are formed during very rapid cooling of a melt in which there is no time for the constituent atoms to arrange themselves into an orderly atomic lattice.

Graphite
A lustrous soft form of carbon usually forming thin layers; often found as nodular inclusions in iron meteorites.

H-chondrite
A group of chondritic stony meteorites belonging to the ordinary chondrite clan. They have the highest total iron of the ordinary chondrites.

HED
Howardites-Eucrites-Diogenites, related basaltic meteorites believed to originate on the asteroid 4 Vesta.

Hexahedrite
An iron meteorite containing less than 5% nickel; usually occurs as nearly pure kamacite with Neumann lines running across the crystal faces.

Howardite
A brecciated achondrite composed of eucrite and diogenite fragments. Thought to be the soil of an asteroid parent body.

Hypersthene chondrites
An obsolete term for meteorites now referred to as L-chondrites.

Hypersthene
A magnesium-rich orthopyroxene $(Mg,Fe)SiO_3$; a common pyroxene in chondritic meteorites.

Individual
A single meteorite individual which has not been fragmented off a larger meteoroid having reached Earth intact.

Interplanetary dust particles
Micron-sized dust particles usually of chondritic composition that are ubiquitous along the Solar System plane and are thought to originate from comets and/or debris from asteroid fragments.

Interstellar grains
Submicron-sized solid grains thought to be ejected by red giant stars. The main constituents are carbon(diamonds) silicon carbide and graphite.

L chondrites
A group within the ordinary chondrite clan containing metal and combined iron in amounts intermediate between the H and LL chondrites.

Lherzolitic
Ultra mafic plutonic igneous rocks composed primarily of olivine and orthopyroxene. They are frequently found with orthopyroxene and clinopyroxene as xenoliths in alkali basaltic rock.

LL chondrites
A group within the ordinary chondrite clan that contains the lowest amounts of metal and combined iron.

Lodranite
A primitive achondrite. Like acapulcoites, they have suffered partial melting in their past history.

Mafic minerals
Silicate minerals that are rich in magnesium and/or iron. These ferromagnesium minerals form mafic igneous rock.

Magma
Molten rock containing dissolved gases and mineral crystals. Through fractional crystallization processes, minerals crystallize out of the magma and interlock to form igneous rock.

Mare basalt
Basalt that forms the maria, the great basins on the Moon. Mare basaltic meteorites are the rarest lunar meteorites.

Matrix
Fine-grained material between inclusions, chondrules and chondrites. This material usually has a similar composition as the chondrules themselves, primarily magnesium-rich olivine and pyroxene.

Mesosiderite
A class of stony-iron meteorites consisting of a mixture of iron/nickel metal and broken rock fragments of magnesium-rich silicate minerals. The rock fragments are similar in composition to eucrites and diogenites of the HED series.

Mesostasis
The last material to solidify from a melt. It is usually found as interstitial fine-grained material or glass between crystalline minerals in an igneous rock.

Metachondrite
A metamorphosed chondrite derived from a chondritic precursor.

Monomict breccia
A brecciated monomict chondrite is composed of angular fragments and matrix all of like composition.

Mosaicism
A characteristic of a mineral crystal seen microscopically under crossed polarizers in which extinction is not uniform but checkered into a mosaic pattern due to small irregularities within the crystal. This occurs when the crystal has been distorted by shock metamorphism. Mosaicism is an indicator of shock effects produced by an impacting body.

Nakhlite
A meteorite from Mars, one of the SNC group. It is composed of the clinopyroxene augite and olivine.

NEA
Near Earth Asteroid. Asteroids whose orbits bring them close to Earth's orbit.

Neumann bands
A network of lines (twinning planes) in an iron-nickel alloy often seen after light chemical etching. Due to mild shock.

Octahedrite
An iron meteorite of intermediate nickel content composed of low nickel kamacite bounded by high nickel taenite arranged in plates on the faces of an octahedrite. Acid etching reveals the Widmanstätten structure.

Olivine bronzite chondrites
This is an obsolete term for meteorites now referred to as H-chondrites, with the "H" standing for high iron. Total elemental iron is between 15 and 19 wt% for the H group.

Paired meteorites
Meteorites that have fallen simultaneously some distance from each other but that are found through analysis to be fragments of the same mass and therefore considered the same meteorite.

Parent body
An astronomical body of subplanetary to planetary size that fragments by collision to produce meteorites.

Parsec
A unit of distance commonly used by astronomers. The distance of an object that would have a stellar parallax of one arc second. 1 parsec = 3.26 light years.

Penetration hole
A hole made by a meteorite that impacts Earth but does not explode.

Perihelion
A position on an elliptical orbit where a celestial body is closest to the Sun.

Pallasite
A class of stony-iron meteorite containing approximately equal amounts of metal and olivine; the metal makes up a continuous network with isolated grains of olivine.

Petrologic type
A scale used to denote the texture of chondritic meteorites; denotes increasing metamorphism in chondrites.

Planetesimal
Small bodies up to a few hundred miles in diameter that formed from the first solid grains to condense out of the solar nebula; planetesimals accreted to each other to form the planets of the solar system.

Plessite
A fine-grained mixture of kamacite and taenite formed late in the diffusion process at low temperatures; usually found filling spaces between the Widmanstätten figures in octahedrite meteorites.

Plutonic rock
Any deep-seated massive body of igneous rock formed beneath the Earth's surface by the consolidation of magma.

Poikilitic texture
A rock texture in which small euhedral mineral grains are scattered without a common orientation within larger typically anhedral grains of different composition. For chondritic meteorites, poikilitic texture is often seen with small olivine grains enclosed in orthopyroxene (see text).

Polymict breccia
A rock made of angular fragments from other rocks of different compositions.

Polymorph
A mineral that has several crystal forms; i.e. graphite is the amorphous form, while diamond is the crystal form, both being polymorphs of carbon.

Porosity
A percentage of the bulk volume of a rock occupied by pore space.

Radiant point
A point in the sky in a specific constellation from which meteors of a meteor shower seem to diverge; an illusion of perspective.

Refractory elements
The first minerals to condense out of a cooling gas at relatively high temperatures; elements that have high vaporization temperatures.

Regmaglypts
Thumbprint-like deep pits or cavities on the exterior of some meteorites produced by the uneven flow of air during the meteorite's passage through Earth's atmosphere. The well-defined polygonal depressions on the surface of these meteorites are ablation features produced during the melting phase of the meteorite's atmospheric passage.

Regolith breccia
A chondritic brecciated meteorite made of consolidated lithified regolith material. Such meteorites have dark/light texture and represent material from beneath the surface as well as on the surface of its parent body.

Retardation point
The point in a meteoroid's path through Earth's atmosphere where its cosmic velocity has dropped to zero and the meteoroid falls freely by Earth's gravity alone.

Rumuruti (R) chondrites
A small clan of meteorites similar to ordinary chondrites but much more oxidized; little if any metal exists, the metal being incorporated into the minerals.

Secondary characteristics
Characteristics resulting from thermal metamorphism, partial melting and aqueous alteration reducing the parent body's primary characteristics to secondary characteristics. They define the physical and chemical history of the meteorite's parent body after its origin but before breakup.

Shergottite
A meteorite from Mars, one of the SNC group. Shergottites are basaltic, with pyroxene, plagioclase and maskelynite as major components.

Siderolite
A meteorite composed of iron and stone, a stony-iron meteorite.

Siderophile elements
The geochemical class of elements with an affinity for the metallic phase rather than the silicate or sulfur phases. Fe, Ni, Co, Cu, Pt (platinum group of metals).

SNC meteorites
An acronym for shergottites, nakhlites, and chassignites. All three are rare achondrite meteorites with young isotopic ages (~1.3 billion years) and are thought to have originated on Mars.

Space weathering
Changes in the spectral properties of surface minerals on asteroidal bodies due to impacts by solar wind particles and micrometeorites. Space weathering may disguise the true spectral characteristics of an asteroid.

Sporadic meteors
An unpredictable, isolated meteor not associated with a periodic meteor shower.

Stony-iron meteorites
A class of meteorite that contain both silicate minerals and iron-nickel metal in approximately equal proportions. Examples: pallasites; mesosiderites.

Terminal velocity
The velocity of a freely falling meteoroid due to Earth's gravity after its cosmic velocity has been reduced to between 320 and 640 km/h. This often marks the end of the visible trail of a fireball.

Tertiary characteristics
Characteristics produced by fragmentation of a meteorite's parent body, shock metamorphism, atmospheric ablation, impact and terrestrial weathering.

Thermal metamorphism
Changes in the chemical and physical characteristics due to internal heating of the parent body, probably by the decay of aluminum-26. Melting temperatures are not reached so that all changes are in the solid state. This is responsible for the various petrographic types of ordinary and carbonaceous chondrites.

Thin section
A slice of rock or mineral that has been ground to a thickness of 0.03 mm and mounted as a slide in a petrographic microscope.

Ultramafic rock
An igneous rock composed of more than 90% mafic minerals.

Ureilite
A rare type of achondrite meteorite consisting of pyroxene grains and olivine set in a carbon rich matrix.

Vitrification
The conversion of a glass to a crystalline texture while in the solid state.

Volatile elements
These are elements that are the last to condense out of a cooling gas. Volatile elements condense from a gas or evaporate from a solid at low temperatures relative to refractory elements. Volatiles are the first materials to be lost when a meteorite is heated.

Widmanstätten structures
An intergrowth of low nickel kamacite and high nickel taenite that mutually grow on the crystal faces of octahedrite meteorites.

Winonaite
A rare class of primitive achondrites that have been partially melted and differentiated. They are associated with IAB irons.

Xenolith
A foreign inclusion in an igneous rock that is not chemically related to the host rock.

Zodiacal light
Light from the Sun that is scattered by interplanetary dust particles along the ecliptic plane and between Earth and the Sun.

Meteorite Index

Italics indicate illustrations

General Index

Italics indicate illustrations

Printing: Ten Brink, Meppel, The Netherlands
Binding: Stürtz, Würzburg, Germany